Early Childhood Teachers' Professi Competence in Mathematics

This edited volume presents cutting-edge research on the professional competence of early childhood mathematics teachers. It considers professional knowledge, motivational-affective dispositions, skills and performance in early childhood mathematics and outlines future fields of research in this area.

The book argues that it is essential for early childhood teachers to prepare a high-quality learning environment and that mathematical competence is highly relevant for children's individual development. Bringing together research from mathematics education, educational science and psychology, it integrates international perspectives and considers the contextual factors that affect the development of children's mathematical competence within Early Childhood Education and Care (ECEC) settings. The book uses a model to describe professional teacher competence that considers the dispositions of early childhood teachers, situation-specific skills of early childhood teachers and the performance of early childhood teachers.

The book is the first of its kind to give a comprehensive overview and allows for integrative perspectives and interdisciplinary understanding regarding pre- and in-service ECEC teachers' professional competence in the domain of mathematics. It will be essential reading for academics, researchers and students of early childhood education, mathematics education and teacher education.

Simone Dunekacke is Assistant Professor for Research on Early Childhood Education at the Freie Universität Berlin, Germany.

Aljoscha Jegodtka is Professor for Social Work at the International University Berlin, Germany.

Thomas Koinzer is Professor for Educational Science at the Humboldt-Universität zu Berlin, Germany.

Katja Eilerts is Professor for Primary School Education in Mathematics at the Humboldt-Universität zu Berlin, Germany.

Lars Jenßen is a Researcher in Educational Science and Psychology with a focus on Mathematics at Humboldt-Universität zu Berlin, Germany.

Routledge Research in Early Childhood Education

This series provides a platform for researchers to present their latest research and discuss key issues in Early Childhood Education.

Books in the series include:

For more information about this series, please visit: www.routledge.com/education/series/RRECE

Early Childhood Teachers' Professional Competence in Mathematics

Edited by
Simone Dunekacke, Aljoscha Jegodtka,
Thomas Koinzer, Katja Eilerts and
Lars Jenßen

Routledge
Taylor & Francis Group
LONDON AND NEW YORK

First published 2022
by Routledge
2 Park Square, Milton Park, Abingdon, Oxon OX14 4RN

and by Routledge
605 Third Avenue, New York, NY 10158

Routledge is an imprint of the Taylor & Francis Group, an informa business

British Library Cataloguing-in-Publication Data
A catalogue record for this book is available from the British Library

Library of Congress Cataloging-in-Publication Data
A catalog record has been requested for this book

ISBN: 9781032000541 (hbk)
ISBN: 9781032000558 (pbk)
ISBN: 9781003172529 (ebk)

DOI: 10.4324/9781003172529

Typeset in Galliard
by KnowledgeWorks Global Ltd.

Contents

Figures

Tables

List of contributors

Camilla Björklund – University of Gothenburg, Sweden

Sigrid Blömeke – Centre for Educational Measurement Oslo, Norway

Esther Brunner – Thurgau University of Teacher Education, Kreuzlingen, Switzerland

Douglas H. Clements – University of Denver, USA

Audrey Cooke – Curtin University, Australia

Febe Demedts – Center for Instructional Psychology and Technology, KU Leuven, Belgium, and ITEC, imec research group at KU Leuven, Belgium

Fien Depaepe – Center for Instructional Psychology and Technology, KU Leuven, Belgium, and ITEC, imec research group at KU Leuven, Belgium

Simone Dunekacke – Free University Berlin, Germany

Lars Eichen – University of Graz, Austria

Katja Eilerts – Humboldt University Berlin, Germany

Susanne Kuratli Geeler – St.Gallen University of Teacher Education, Switzerland

Herbert P. Ginsburg – Teachers College, Columbia University, USA

Shannon Stark Guss – University of Denver, USA

Aiso Heinze – Leibniz Institute for Science and Mathematics Education, Germany

Georg Hosoya – Freie Universität Berlin, Germany

Jenny Jay – Curtin University, Australia

Aljoscha Jegodtka – IUBH University of Applied Sciences, Germany

Lars Jenßen – Humboldt University Berlin, Germany

Thomas Koinzer – Humboldt University Berlin, Germany

Miriam Leuchter – University of Koblenz Landau, Germany

Anke Lindmeier – University of Jena, Germany

Anuschka Meier – University of Teacher Education in Special Needs, Switzerland

Elisabeth Moser Opitz – University of Zurich, Switzerland

Hanna Palmér – Linnaeus University, Sweden

Manfred Pfiffner – Zurich University of Teacher Education, Switzerland

Lara Pohle – Humboldt University Berlin, Germany

Karoline Rettenbacher – University of Graz, Austria

Julie Sarama – University of Denver, USA

Corinna Schmude – Alice Salomon Hochschule Berlin, Germany

Selma Seemann – Leibniz Institute for Science and Mathematics Education, Germany

Markus Szczesny – Technische Universität Braunschweig, Germany

Oliver Thiel – Queen Maud University College of Early Childhood Education, Norway

Joke Torbeyns – Center for Instructional Psychology and Technology, KU Leuven, Belgium

Franziska Vogt – St.Gallen University of Teacher Education, Switzerland

Catherine Walter-Laager – University of Graz, Austria

Andrea Wullschleger – University of Zurich, Switzerland

Introduction

Simone Dunekacke, Aljoscha Jegodtka,
Thomas Koinzer, Katja Eilerts, Lars Jenßen

Supporting mathematical competence has been identified as an important issue in early childhood education in order to ensure successful participation in knowledge-based societies around the world (Early et al., 2007; OECD, 2018). A growing body of research in recent years has addressed pre- and elementary school children's development of mathematical competence (Linder & Simpson, 2018). This research has highlighted that early mathematical competence is highly important for the development of mathematical competence in later years at school (Duncan et al., 2007; Krajewski & Schneider, 2009; Litkowski, Duncan, Logan, & Purpura, 2020; Nguyen et al., 2016). The development of mathematical competence in children's early years depends on several aspects, ranging from individual characteristics (e.g. children's general cognitive abilities) to environmental ones, such as parents' socioeconomic status or specific learning opportunities provided by the parental home or in institutional early childhood education and care (ECEC) settings (Burghardt, Linberg, Lehrl, & Konrad-Ristau, 2020; Bronfenbrenner & Morris, 2007). Institutional ECEC settings have been identified as relevant learning environments for mathematical competence, in particular (Litkowski et al., 2020; Ulferts & Anders, 2016). However, in order to experience these advantages, ECEC institutions have to offer children high-quality learning opportunities. High-quality learning opportunities in the field of early mathematical learning are characterized by a focus on children's needs and interests as well as on core mathematical ideas in terms of content (Clements & Sarama, 2016; Gasteiger & Benz, 2018; van Oers, 2010). The quality of such learning opportunities depends to a great extent on ECEC teachers' professional competence. Research has shown that domain-specific aspects, for example with respect to mathematics (McCray, 2008), play a crucial role in addition to general aspects of teachers' professional competence (Parks & Wager, 2015). Domain-specific aspects include, for example, knowledge about children's domain-specific development or domain-specific aspects of learning support. ECEC teachers' professional competence in mathematics has been the subject of a growing body of research in recent years. However, there is still a lack of systematic

DOI: 10.4324/9781003172529-101

literature in this field (Linder & Simpson, 2018). Specifically, no comprehensive overview of descriptions, development and effects of ECEC teachers' competence in mathematics currently exists. This edited volume serves as a first step to closing this literature gap.

While the claim that early mathematical learning should address children's needs as well as core mathematical ideas may appear obvious, it might be challenging to implement and rely on specific features of educational contexts as pre- and elementary-school settings. Moreover, several approaches and opportunities to foster early mathematical learning exist. These approaches range from more social pedagogical to more curriculum-based approaches. Social pedagogical approaches tend to use everyday activities and free play to foster children's (mathematical) learning. Consequently, learning opportunities are more random and must be identified by the ECEC teachers as part of the activity. In contrast, curriculum-based approaches offer more structured, prepared opportunities for children's learning (Clements & Sarama, 2016). These two approaches are associated with heterogeneous requirements for EC teachers' professional competence. First, in both approaches, ECEC teachers have to identify children's level of mathematical competence. Second, in both approaches, ECEC teachers must interact with the children, for example by employing sustained shared thinking (Siraj-Blatchford, Sylva, Muttock, Gilden, & Bell, 2002), while supporting them. In addition, within these approaches, ECEC teachers are required to prepare interesting opportunities to learn that are also appropriate for the children's developmental level. In the social pedagogical approach, ECEC teachers have to identify opportunities to learn during various everyday activities, whereas in curriculum-based approaches, ECEC teachers might engage in planned learning opportunities.

In addition to these differences in setting and requirements, differences regarding ECEC teacher education also exist. It is recognized that ECEC teachers need to possess domain-specific competency opportunities to learn during initial teacher training and professional development in order to develop these competences (Blömeke, Jenßen, Grassmann, Dunekacke, & Wedekind, 2017; Clements & Sarama, 2016; Ginsburg, 2018). When considering the formal qualification needed to work in ECEC settings, both countries involving non-university ECEC teacher education as well as countries where the minimum qualification of an ECEC teacher is a bachelor's or master's degree can be observed (European Commission, EACEA, Eurydice, & Eurostat, 2014; Gasteiger, Brunner, & Chen, 2020; Oberhuemer, Schreyer, & Neuman, 2010). More specifically, mathematics-related learning opportunities in ECEC teacher education differ both between countries (Gasteiger et al., 2020; Oberhuemer et al., 2010) and within countries (e.g. Germany: Blömeke et al., 2017). Finally, these differences in pedagogical approaches and in ECEC teacher education systems are reflected in different terms for the people who work with children prior to school entry. The term *ECEC educator* is often used in more social pedagogical approaches, while the term *ECEC teacher* is often used in more curriculum-based approaches. Readers should take this into account when reading the contributions making up this edited volume. The authors of each chapter used the term most suitable for their country or their research.

While supporting young children's mathematical learning is a common requirement across pedagogical approaches and teacher education systems, this edited volume applies the general framework of teacher competence by Blömeke, Gustafsson, and Shavelson (2015) to structure and locate the individual contributions regarding ECEC teachers' professional competence. Professional competence in the framework by Blömeke and colleagues (2015) is conceptualized as a conglomerate of different constructs (see Figure 0.1). *Cognitive and affective-motivational aspects* represent the dispositional constructs of professional competence. According to Shulman's framework (1986) and its applications to mathematics (Hill, Rowan, & Ball, 2005), the cognitive construct in the field of mathematics education can be differentiated into mathematical content knowledge (MCK), mathematical pedagogical content knowledge (MPCK), and general pedagogical knowledge (GPK). Emotions, such as mathematics anxiety or enjoyment, and beliefs, such as mathematics self-efficacy, are understood as affective-motivational dispositions. *Situation-specific skills* are considered another relevant dimension of professional competence. In early childhood mathematics education, the *perception* of math-related issues, *interpretation* of children's mathematical performance, and skill in *planning* math-related activities are described as important situation-specific skills. It is theoretically assumed that situation-specific skills mediate the relationship between cognitive and affective-motivational dimensions and ECEC teachers' performance. From a practical point of view, ECEC teachers' *performance* is the crucial

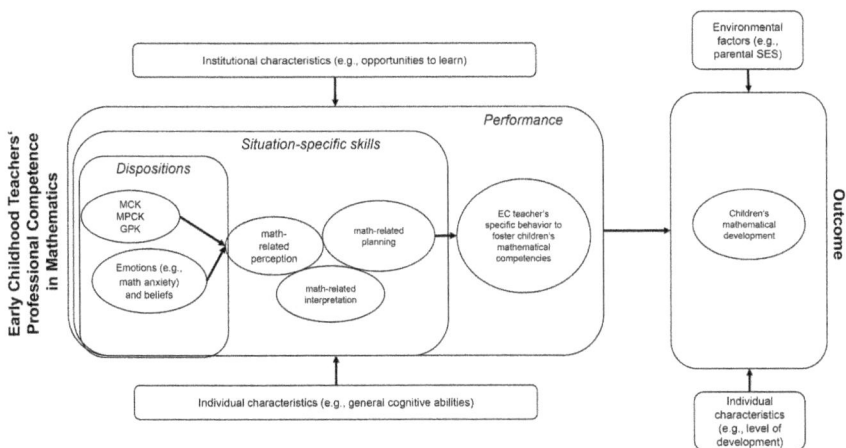

Figure 0.1 Framework of professional competence.

dimension of interest. Within this framework, performance is defined as ECEC teachers' observable behaviour, such as concrete math-related interactions with the children. Performance is the dimension that links teachers' professional competence to children's learning.

This framework of professional competence will not only be addressed in the contributions making up the edited volume but also represented in the structure of the book itself. The book is structured into three main parts:

I   Dispositions of early childhood teachers' professional competence
II Situation-specific skills of early childhood teachers
III Performance of early childhood teachers

Part I starts with an essay by Herbert Ginsburg reflecting on learning activities during a mathematics course within a teacher training program. Joke Torbeyns, Febe Demedts, and Fien Depaepe present research regarding the development of MPCK during teacher training. In-service teachers' performance expectations regarding the children and learning objectives, which are important aspects of ECEC teachers' MPCK, are addressed in the research presented by Karoline Rettenbacher, Lars Eichen, Manfred Pfiffner, and Catherine Walter-Laager. Oliver Thiel addresses emotions about mathematics and their development during teacher training. Emotions regarding mathematics are also addressed in the chapter by Franziska Vogt, Miriam Leuchter, Susanne Kuratli Geeler, Simone Dunekacke, Aiso Heinze, Anke Lindmeier, Anuschka Meier Wyder, Elisabeth Moser Opitz, Selma Seemann, and Andrea Wullschleger, with a focus on in-service teachers. Part I closes with a systematic review by Lars Jenßen addressing pre- and in-service ECEC teachers' mathematics anxiety.

Part II addresses ECEC teachers' situation-specific skills. Audrey Cooke and Jenny Jay discuss how Bishop's fundamental ideas about mathematical activities can be used to support ECEC teachers in mathematics activities with preverbal children. Simone Dunekacke and Sigrid Blömeke report on a study on the development of situation-specific skills in ECEC teacher education.

Part III addresses ECEC teachers' performance in early mathematics education. The research by Julie Sarama, Douglas H. Clements, and Shannon Stark Guss shows how a professional development programme based on learning trajectories can enhance teachers' performance and children's learning. Camilla Björklund und Hanna Palmér present research on how ECEC teachers and children use mathematics-related picture books to support children's mathematical learning. The relationship between children's mathematical competence and the extent of math-related talk in circle time provided by ECEC teachers is investigated by Aljoscha Jegodtka, Georg Hosoya, Markus Szczesny, Lars Jenßen, and Corinna Schmude. Finally, Lara Pohle, Lars Jenßen, and Katja Eilerts investigate what kind of mathematical activities ECEC teachers offer to children.

References

Blömeke, S., Gustafsson, J.-E., & Shavelson, R. J. (2015). Beyond dichotomies: Competence viewed as a continuum. *Zeitschrift für Psychologie, 223*(1), 3–13. https://doi.org/10.1027/2151-2604/a000194

Blömeke, S., Jenßen, L., Grassmann, M., Dunekacke, S., & Wedekind, H. (2017). Process mediates structure: The relation between preschool teacher education and preschool teachers' knowledge. *Journal of Educational Psychology, 109*(3), 338–354. https://doi.org/10.1037/edu0000147

Bronfenbrenner, U., & Morris, P. A. (2007). The bioecological model of human development. In W. Damon & R. M. Lerner (Eds.), *Handbook of child psychology* (pp. 793–828). Hoboken, NJ: John Wiley & Sons, Inc. https://doi.org/10.1002/9780470147658.chpsy0114

Burghardt, L., Linberg, A., Lehrl, S., & Konrad-Ristau, K. (2020). The relevance of the early years home and institutional learning environments for early mathematical competencies. *Journal for Educational Research Online, 12*(3), 103–125.

Clements, D. H., & Sarama, J. (2016). Math, science, and technology in the early grades. *The Future of Children, 26*(2), 75–94.

Duncan, G. J., Dowsett, C. J., Claessens, A., Magnuson, K., Huston, A. C., Klebanov, P., Pagani, L. S., Feinstein, L. Engel, M., Brooks-Gunn, J., Sexton, H., Duckworth, K., & Japel, C. (2007). School readiness and later achievement. *Developmental Psychology, 43*(6), 1428–1446. https://doi.org/10.1037/0012-1649.43.6.1428

Early, D. M., Maxwell, K. L., Burchinal, M., Alva, S., Bender, R. H., Bryant, D., Cai, K.; Clifford, R. M., Ebanks, C., Griffin, J. A., Henry, G. T., Howes, C., Iriondo-Perez, J., Jeon, H.-J., Mashburn, A. J., Peisner-Feinberg, E., Pianta, R. C., Vandergrift, N., & Zill, N. (2007). Teachers' education, classroom quality, and young children's academic skills: Results from seven studies of preschool programs. *Child Development, 78*(2), 558–580. https://doi.org/10.1111/j.1467-8624.2007.01014.x

European Commission, EACEA, Eurydice, & Eurostat (2014). *Key data on early childhood education and care in Europe* (2014 Ed.). Eurydice and Eurostat report. Luxembourg: Publications Office of the European Union.

Gasteiger, H., & Benz, C. (2018). Enhancing and analyzing kindergarten teachers' professional knowledge for early mathematics education. *The Journal of Mathematical Behavior, 51*, 109–117. https://doi.org/10.1016/j.jmathb.2018.01.002

Gasteiger, H., Brunner, E., & Chen, C.-S. (2020). Basic conditions of early mathematics education – A comparison between Germany, Taiwan and Switzerland. *International Journal of Science and Mathematics Education, 16*(3). https://doi.org/10.1007/s10763-019-10044-x

Ginsburg, H. P. (2018). Helping teacher educators in institutions of higher learning to prepare prospective and practicing teachers to teach mathematics to young children. In G. Kaiser, H. Forgasz, M. Graven, A. Kuzniak, E. Simmt, & B. Xu (Eds.), *ICME-13 monographs. Invited lectures from the 13th International Congress on Mathematical Education* (1st ed., pp. 135–154). Cham: Springer International Publishing. https://doi.org/10.1007/978-3-319-72170-5_9

Hill, H. C., Rowan, B., & Ball, D. L. (2005). Effects of teachers' mathematical knowledge for teaching on student achievement. *American Educational Research Journal, 42*(2), 371–406. https://doi.org/10.3102/00028312042002371

Krajewski, K., & Schneider, W. (2009). Early development of quantity to number-word linkage as a precursor of mathematical school achievement and mathematical difficulties: Findings from a four-year longitudinal study. *Learning and Instruction, 19*(6), 513–526. https://doi.org/10.1016/j.learninstruc.2008.10.002

Linder, S. M., & Simpson, A. (2018). Towards an understanding of early childhood mathematics education: A systematic review of the literature focusing on practicing and prospective teachers. *Contemporary Issues in Early Childhood, 19*(3), 274–296. https://doi.org/10.1177/1463949117719553

Litkowski, E. C., Duncan, R. J., Logan, J. A. R., & Purpura, D. J. (2020). When do preschoolers learn specific mathematics skills? Mapping the development of early numeracy knowledge. *Journal of Experimental Child Psychology, 195*, 104846. https://doi.org/10.1016/j.jecp.2020.104846

McCray, J. S. (2008). *Pedagogical content knowledge for preschool mathematics: Relationships to teaching practices and child outcomes.* Chicago: ProQuest.

Nguyen, T., Watts, T. W., Duncan, G. J., Clements, D. H., Sarama, J. S., Wolfe, C., & Spitler, M. E. (2016). Which preschool mathematics competencies are most predictive of fifth grade achievement? *Early Childhood Research Quarterly, 36*, 550–560. https://doi.org/10.1016/j.ecresq.2016.02.003

Oberhuemer, P., Schreyer, I., & Neuman, M. J. (2010). *Professionals in early childhood education and care systems: European profiles and perspectives.* Opladen: Budrich.

OECD (2018). Early learning matters: The international learning and well-being study. http://www.oecd.org/education/school/Early-Learning-Matters-Project-Brochure.pdf accessed 25.01.2021.

Parks, A. N., & Wager, A. A. (2015). What knowledge is shaping teacher preparation in early childhood mathematics? *Journal of Early Childhood Teacher Education, 36*(2), 124–141. https://doi.org/10.1080/10901027.2015.1030520

Shulman, L. S. (1986). Those who understand: Knowledge growth in teaching. *Educational Researcher, 15*(2), 4–14.

Siraj-Blatchford, I., Sylva, K., Muttock, S., Gilden, R., & Bell, D. (2002). *Researching effective pedagogy in the early years. Research report: Vol. 356.* Nottingham: Department for Education and Skills.

Ulferts, H., & Anders, Y. (2016). *Effects of ECEC on academic outcomes in literacy and mathematics: Meta-analysis of European longitudinal studies.* http://ecec-care.org/fileadmin/careproject/Publications/reports/CARE_WP4_D4_2_Metaanalysis_public.pdf (accessed 25.01.2021).

van Oers, B. (2010). Emergent mathematical thinking in the context of play. *Educational Studies in Mathematics, 74*(1), 23–37. https://doi.org/10.1007/s10649-009-9225-x

Part I

Dispositions

Chapter 1

Voices of competence

What I learned from my early education students

Herbert P. Ginsburg

Introduction

For many years, I taught a graduate-level course, *The Development of Mathematical Thinking*, for prospective early childhood teachers at Teachers College, Columbia University. This chapter is a reflection on my personal experience in teaching that course. I describe what I learned from my students' reflective voices about their competence: their fears of math; their attitudes toward teaching math to young children; their ideas and knowledge about how young children think and learn and how teachers should teach; and their skills of observation, interviewing, and interpretation of children's behavior.

This chapter is also about what I learned from my students, particularly the vital pedagogical role of videos depicting children's learning and thinking, the benefits of the student interview assignment, the unanticipated power of the student "reflection," and the need for close links with the schools.

Background

I write here about two of my classes, one in the spring of 2016, with 20 students, and one in the fall of 2016, with 13 students.

Almost all of my students were female early childhood education majors, at the Masters level. The course was required; I suspect that if it were not, many of these students would not enroll in it.

The course involved 15 sessions, once a week, for an hour and 40 minutes, in the late afternoon—that is, after some of the students had been working all day at their jobs, often in classrooms. Each week there were assigned readings and videos available on the class website, a lecture, class discussions, and various assignments. The course required a mid-term project and a final project, but no formal examinations.

The course focused heavily on children's mathematical thinking. The content comprised children's informal mathematical knowledge in the everyday environment, from the age of 2 or so onward; the learning of formal, school mathematics, beginning with the introduction of symbolism in Kindergarten and first

DOI: 10.4324/9781003172529-1

grade; the roles and nature of memory, strategy, and understanding in school math from Kindergarten to about fourth grade; the math content, particularly number, operations on number, pattern, shape, space, and measurement; the assessment methods of testing, observation, and clinical interview; the processes of teaching; and the roles of free play, curriculum, manipulatives, and computer software in children's math learning.

After each class, I required students to submit on the class website a reflection on the latest session. Most often, I asked general questions: "What interested you the most? What did you learn? What was arguable?" Sometimes, I asked focused questions: "How do you feel about teaching math?" "What is the role of free play?" I encouraged the students to submit their reflections as soon as possible, while the material was fresh in mind. I told the students that I would respond to, but not grade, each of their reflections. I assured them that other students would not see their reflections or my responses to them and that at the outset of the next class, I would discuss some of their reflections.

Next, I describe what I learned from the students' weekly reflections about fears of math, the issue of "math is all around us," everyday activities and free play, math in storybooks, the nature of the child's mind, and the benefits of the clinical interview. There were more topics about which I could have written, but these, I think, were the most important, showing how teacher competence is complex indeed (Blömeke, Gustafsson, & Shavelson, 2015).

Weekly student reflections

I found the reflections to be enormously informative. I greatly enjoyed reading them, commenting on them, and discussing some at the outset of the next lecture. Over time, the students revealed their deepest thoughts and feelings about early math education and the issues under discussion, and I responded honestly to each student's refection each week. The impersonality of the digital environment curiously afforded a kind of shared intimacy that might not have been possible in a live class and allowed students to reveal sensitive thoughts.

These reflections—the students' voices—provide the foundation for this chapter. I have divided their reflections into several topics, beginning with student fears. Note that in all cases, the indented reflections are literal accounts of what the students wrote.

Fears

I expected that many students would describe their fears of math and teaching it. Many—but not all—did, but I was surprised by one aspect of their reactions.

> Overall, this week's lecture made me realize that I am not alone in my feelings of discomfort surrounding teaching math in the classroom.

**

> I was pleasantly surprised by the students in today's class. I was concerned that I was going to be the only person that wasn't good at math coming into the course. Yet that wasn't true!

It hadn't occurred to me that students felt alone in their negative feelings about math. I knew from my teaching experience and from the research (Herts, Beilock, & Levine, 2019) that math anxiety is widespread but did not appreciate the sense of isolation that at least some students associate with it.

Some students speculated on the origins of those fears.

> When my math teacher focused on teaching "math skills", which is using certain formula to solve certain types of math problems without understanding the concept of the math, and when math teacher judged my ability only based on my math grades, it was hard for me to love math. However, when my math teacher spent enough time to explain new math concepts and respected different learning styles or speeds, math became a very enjoyable subject.

> **

> I thought back to when my math anxiety was formed. I thought back specifically to one class. I remember the teacher seemed so unsure of how to help me, which led to her just passing over my questions. Her unsure composure led to me being completely unsure of how I was learning and how I was unable to grasp the concepts. That anxiety was passed down from my teacher to me.

These recollections suggest that at least some children do not begin schooling with math anxiety; instead, it may result from teachers' behavior and attitudes. If so, then it is incumbent on prospective teachers to overcome their anxieties and learn to teach in ways that make math an exciting and enjoyable subject to learn.

Here's another response that surprised me.

> Most people stated that they were not great at math and that they were excited to learn how to teach it, just like I was.

Apparently, despite this student's belief that she was not "great at math," and probably had negative feelings about it, she—like others in the class—was eager to learn about early math education. What more could I ask for?

In brief, many students believe they are alone in their math anxiety; some trace it back to their teachers; and despite it, some are intent on learning to teach math.

Is math really all around us?

Most early childhood educators want students to understand that "math is all around us" and that free play is an important contributor to math learning (Singer, Golinkoff, & Hirsh-Pasek, 2006). In one lecture, I began with a simple exercise in which students were asked to identify math in various everyday scenes—pictures of animals, of a preschool classroom, of clothing. Here are several student reflections:

> With the first photo of flowers, I instantly only thought about counting the flowers. But then as we continued to discuss it as a class I noticed other things such as the leaf with the line down the middle which splits it in half. There were so many different answers, some of my classmates said things I hadn't even thought of.

**

> Looking at everyday scenarios and picking out different math concepts that could be taken from such scenes was especially compelling to me. My mind has definitely been opened to the idea that "math is all around us," to the point where I already found myself thinking with this new mindset. It was kind of hard to turn off thinking that way **even though I was supposed to be looking for literacy** [my emphasis] in the classroom.

Clearly, my students were captivated by the idea that "picking out different math concepts that could be taken from such scenes was especially compelling ...." This is a complex idea that bears discussion. Sometimes people say that "Math is all around us." In one sense, this is not true. The picture I showed of the flower, for example, does not contain any math. A flower is a flower. The math we attribute to the flower is in our minds. We conceptualize a "line down the middle which splits it in half" or have the idea that there is reflection symmetry in a flower. As the students put it, these "... different math concepts **could** [my emphasis] be taken from such scenes ...." Math is ubiquitous only if we are equipped and motivated to see it. Prospective teachers, therefore, need to be helped first to understand the math itself and then learn to look for it in the everyday world, particularly the classroom.

But that's only the first step.

> The most important take away for me ... is that though mathematics is often presented as being all around us, it is found in form of objects and 'mathematizing' these objects is up to us as teachers.

**

> It was interesting to see how you can pretty much "mathematize" anything. Windows can be used for fractions and turned into quadrants ....

**

> Many of the things in the classroom we can use as math for our students, as long as we put meaning to it.

"Mathematizing" means making explicit the math that we see in objects, and as we will see children's behavior as well. The fact that we adults can see math in the design of a dress does not ensure that a casual or even extended look at it will help a child learn about its symmetries. Merely exposing children to a math-rich world is not sufficient to teach children the math we see. Saying "math is all around us" is not a teaching strategy and unfortunately may even serve as an excuse for not teaching math. Teachers have to mathematize what is seen and done; they have to "put meaning to it." They need to engage children in explicit mathematical activity and thinking.

In brief, my students learned to appreciate the math that can be seen (or conceived of) in the everyday environment. They also saw the need to make this math explicit—to mathematize it in ways that can engage children and stimulate their thinking.

Everyday activities and free play

After asking the class to identify math in photos of everyday scenes, I showed a video of two children playing with blocks. They were sitting together at a small table and created various constructions—a castle, a house, a railroad train. The children had not been told to do anything. It was "free play"—their turn to play with the blocks in any way they chose. Block play of this kind is popular in many classrooms. Indeed, it has a long history in early education, and a great deal has been written about the activity (Ginsburg, Inoue, & Seo, 1999; Hirsch, 1996).

I asked my students to look carefully at the video, talk about what they observed, and interpret what they saw. We discussed how the children clearly knew the differences among the various shapes, could combine them to create other shapes, and could construct symmetrical buildings, like a castle with two towers, each with the same shape on both sides. It was very clear that children's block play involves some implicit and often non-verbalized mathematical ideas about shape, size, position, up and down, in front and behind, and symmetry. Some of the ideas are "embodied" (Lozada & Carro, 2016) in the sense of being tied to the child's actions, like *reaching over* some blocks to put something *behind* them. The block constructions also reveal an aesthetic sensibility.

One student reflected on the process of analyzing the video in class.

> I found it remarkable how much can be learned from just watching the first few seconds of a recording! I cannot stop thinking about how much learning goes unnoticed in the early childhood classroom. If only life had a pause and rewind button, then I could effectively analyze the mathematical thinking occurring in the classroom without missing a beat.

I asked the class to reflect on other examples of children's everyday math that they may have witnessed.

> I used to listen to my students talking about their food and I made some questions about it. It was amazing the amount of mathematical thinking they used during those free time (amount of food, comparing the amount of food, how many were left, etc.).

**

> I was surprised to see a 4 year-old child's geometrical thinking when she told me, "You know? I do not use the bottom cup because all the circle of it has touched the table, and it is dirty."

Students offered many more examples of children's everyday math. They were convinced that not only is math all around us (or rather that we can see it if we know how to look) but that children engage in many informal mathematical activities and have an "everyday math" of some complexity.

Some students also reflected on their role as facilitators and mathematizers (a new word!) of children's everyday math. One impetus to these reflections was a paper by Kontos (1999).

> When I heard my students using math concepts during lunch time. I enjoyed sitting down with them and entering to their conversations, in a similar way as Kontos' "play enhancer or playmate" role of teacher.

Finally, it's all well and good for the teacher to attend to the different ways that children spontaneously use math in the everyday world. But in practice, many problems arise. One student wrote that teachers often cannot help individual children during free play:

> ... because they are setting up and preparing for a variety of other things, and thats a little upsetting. But the more I thought about it, I see that quite often thats the truth, its sad but its true. When I think about it most places allow for an hour maybe of "free play" or "work time" and within that hour you want to engage with twenty or so children in a real, meaningful, and appropriately challenging way. And I can see how that may be challenging especially with only one or two teachers and a whole bunch of deadlines and material to cover.

In other words, if you as a teacher are in a room with some 20 children engaged in free play, you need 20 heads and brains to promote each child's learning in a sensitive manner. There is no way that you can respond effectively to each of your different children. But, I told my students when this issue came up in class: don't despair. Yes, responding to each child is impossible for a normal, one-headed, one-brained person, but you can certainly work with some children. Remember too that children sometimes need to be left alone so that they can overcome difficulties and learn on their own. And sometimes children learn from each other. Finally, exploiting free play is only one approach to early mathematics education. It is also necessary to teach, to introduce "curriculum" activities.

In brief, students learned that children's free play and everyday interactions contain a good deal of implicit math that teachers can mathematize, although doing so is not always easy (and is not the only desired approach to early education).

Math in stories

If math is all around us, it is in storybooks too. Concerned and comfortable as they are with literacy, many teachers read stories to their children, at least once a day. Occasionally, teachers read "math books" about counting, shapes, pattern, and more. Also, some picture books not normally considered to be "math books" really are. Indeed, the apparently non-math books sometimes contain richer math content than do the books designed to teach math (Ginsburg, Uscianowski, Carrazza, & Levine, 2020).

A simple example is *The Growing Story*, in which a young boy sees the chicks and the flowers grow as the seasons pass over the course of year. But he cannot see himself grow. When winter arrives, he wonders whether he has grown at all. But when he begins to put on last year's winter outfit .... Well, you have to read the book to find out what happened.

Because I believe that storybook reading can have a major role in promoting math learning, I had my class analyze the math and the illustrations depicting it in various storybooks. The students were very excited. After all, who—including the most math averse among us—does not like reading storybooks with children?

My students' subsequent reflections highlighted several themes. One was to stress that ordinary non-math books may offer rich math material.

> After reading *The Growing Story* in class, I was able to see how much math actually exists in a picture book. From figuring out measurements to simply counting items on a page, math problems were able to be pulled from each page. I was surprised to see how much math children could actually learn from … reading a book.

> **

> The story book we looked at in class, "The Growing Story," is not one that I would have initially chosen to tie in with a math class. Going through it, it was a challenge to connect it to any mathematical thinking, but on deeper analysis there were several ways in which perspective, rate, comparison, and measurement could be integrated into the story. I really look forward to doing more such activities in class, which integrate aspects of both literacy and math.

Other students offered a caveat, namely that the focus on math might be problematic.

> I saw how much can be done if one is purposely looking for ways to encourage mathematical thinking while telling a story. It remains a concern to me, nevertheless, whether children may be distracted or upset by continuous interruptions and inquisitions.

I agree. A teacher can overdo math teaching during book reading. The math is interesting, but above all, the child wants to know: what happens next? What happens when the boy puts on the clothes? So, the problem for the teacher is how to discuss the math without destroying the story.

Finally, one of the reasons I favor use of math storybooks is that they offer teachers averse to math and math teaching a gentle introduction to math education for young children. The books can be fun to read and the teacher can come to learn that the math in the books is not very hard and that teaching it need not be noxious.

> I like the picture book activity most. It turns out that I have taught some math knowledge to my students in a way that I even did not notice.

In brief, the students reflected that the math content pervading picture books, sometimes stealthily, can be used to introduce math ideas in a way that teachers enjoy and children love—through literacy.

Children's mathematical thinking

I devoted many sessions of the course to reviewing the research literature—experimental studies, clinical interviews, and naturalistic observations—on young children's mathematical thinking from about two years to eight years of age (Clements & Sarama, 2014; Ginsburg, 1989). The basic questions were these: first, how do young children understand mathematical ideas and solve informal mathematical problems before written math is formally introduced in the context of schooling? For example, what are children's ideas about adding, and their methods for solving addition problems, before they encounter written addition in school? Second, once they experience formal education, how do children understand school math, with its symbols, algorithms, formulas, and all the rest?

Why focus on student thinking? First, students find it fascinating (as I do), and sometimes even shocking. In the problem this student describes, a boy had five pirate coins and so did the interviewer. Both coin collections were then covered. The interviewer took one from him (now he has four) and put it in her pile (she now has six). She then asked him to make both piles the same number:

> I was fascinated by the little boy's ability to make both pirate coins equal when the researcher had removed one coin from his pile and added it to her pile without him being able to see her total amount. I assumed that he would add one coin back to his. I was shocked that he had the reasoning skills to add one for the one removed, PLUS an additional coin to match the researchers gain.

Another shock was about children's systematic misconceptions. The problem was Piaget's classic conservation of equivalence, in which a child first determines that two lines of toy bears, one carefully placed just below the other, both have seven bears. Then, after one of the rows is spread out, the child counts them again. The question then is whether the two rows still have the same number of bears.

> I was shocked to see the video of Liola, who actually counted both lines of bears but then after the instructor spread out the bears in one of the rows, Liola decided that the longer row had more.

Students also learned that learning math is more complex than they originally thought.

> We may think of a child's counting to be a result of memorization and repetition. However, I learned that there is so much more than just rote learning. Children are in fact, engaged in abstract mathematical thinking as they are counting. It was fascinating to see Anna use her own strategy and knowledge of the units to count all the way up to 100. (Ginsburg, 2016)

A second reason for the interest in children's thinking is that knowledge of it can guide instruction (Carpenter, Franke, Johnson, Turrou, & Wager, 2017).

> I was amazed by young children's thinking process when they are handling math problems. Their intuitive yet effective strategies showed me how brilliant those young minds are and our teaching should adapt to their individual strength and interests instead of failing them.

Some students used what they learned to reevaluate their approach to children.

> We make rapid assumptions that a child must mean X when he/she says Y, but we very seldom let them explain what they truly mean and how they came to the conclusion .... When I speak to children now, I try my best to understand them, or not to jump into conclusions .... I think it's a little weird that I am doing this now because this isn't part of any of my homework assignments.

Some students used what they learned about children's thinking to propose new specific approaches to instruction.

> When I thought of counting before reading this last chapter I only thought about the rote memorization of numbers, as that is how I remember learning to count. But the strategy of helping students get from 29 to 30 for example, by asking what number comes after 2 really stood out to me. [*The student means that if the child knows that 3 comes after 2, then 30 should come after 20*]. I think this strategy is a great one to try and teach children rather than just hinting at the sound of the next number ... By showing children this pattern, the concept of counting may click in their minds. And as the reading pointed out, this strategy may also lead to children understanding place value, even before entering an elementary school classroom!

Students also learned that the math to be taught—even counting—is more complex than ordinarily supposed.

> Before I had never thought that counting involved so many underlying mathematical concepts and rules (cardinality, order irrelevance, etc.) because we adults have practiced it so many times that it almost becomes automatic.

**

> "One is an idea" That phrase is something that stuck with me after reading this paper. "One," to me, has always been such an absolute term, tangible, measurable and finite, that to now think of one as an idea, is quite fascinating to me. This paper also made me think about the fact that a set of two is one, one, a set of three is one, one, one and it is as though the number names are actually just collective nouns for different sets of ones.

In brief, the students learned that children's math thinking is complex. Sometimes children exhibit surprising competence and sometimes a surprising lack of it. Knowledge of children's thinking can guide instruction in general and specific ways and can also help students come to grips with the depth of the math that they are expected to teach.

The clinical interview

The students also learned about assessment—standardized testing, observation, and clinical interview. My main focus was on the clinical interview: asking questions to clarify what children know (Ginsburg, 1997). In almost every class, students studied at least one clinical interview video highlighting children's thinking. As we have seen, students were frequently surprised, delighted, and shocked by what the clinical interview revealed through its flexible, deliberately unstandardized questioning.

Virtually all students saw great value in interviewing.

> This course has definitely opened my eyes to the importance of conducting clinical interviews as often as possible. Simply assessing children through their abilities to produce the correct answer does not represent their full understanding, nor does it show their readiness to progress towards more complicated concepts.

One student mentioned that the child could enjoy the activity.

> The child not only understood the concepts behind what he was doing, but he also enjoyed what he was doing. It was like play for him.

This insight is important because many students assume that children's experience with math is as unpleasant as their own was. On the contrary, engaging in mathematical thinking can be a form of play.

For the mid-term examination, students were required to conduct an interview with an individual child. Students learned that doing so was not easy!

> I learned clinical testing is nowhere near as easy as it looks. When watching an interview, it's so easy to critique what the interviewer did wrong … but when you are actually doing the interview yourself it suddenly doesn't seem so easy anymore. You never know what the child is going to say, and you are so caught up in not trying to say the wrong thing it's possible to lose focus on the task.

Despite difficulties like these, almost all students found great value in the interview experience.

> The level of nuance I got from the interview I feel is much more helpful than some tiered system between kids who are "good at math", kids who are "at the same level as the class," and kids who are "struggling."

Some students noticed that the clinical interview not only reveals children's thinking but can even stimulate it.

> The idea that a clinical interview can change the way a child deals with questions, making her more likely to think and engage in new ways of reasoning, is also very interesting.

Another student went even further in characterizing the value of the interview.

> My favorite part was watching the real-life example of Dr. Ginsburg teaching math to the little girl. This way of teaching allows for the student to figure out his/her wrong answer on their own as opposed to simply being told the right answer or told that they are wrong. This way also seems to avoid discouraging the student from trying to solve the problem again and embarrassing them that they have given a wrong answer. Overall, I thought it was a great approach to teaching at the early childhood level.

I was surprised because the primary goal of my interview was not to teach, but to reveal the child's thinking. I had made a clear distinction between assessment (interviewing to find out what the child knows) and teaching (helping the child to learn). But I think that the student was right. Interviewing can be a combination of both assessment and teaching. For researchers, the conflation of the two can contaminate a controlled study. But for teachers, the mixture is a blessing. The clinical interview can include a subtle form of teaching.

One student summarized what she learned from the clinical interview in a way that I found gratifying. This idea should inspire all teachers, but especially those who work with children who typically fail in school.

> You can't assume that children are not intelligent based on surface knowledge. Rather, it is your job to find the spark in each child, exploring what they themselves can succeed at and are interested in.

Yes, the "spark." Finding it is what we should all aim for, in our own students, and in theirs.

In brief, the students learned that the clinical interview is a powerful, enjoyable, and difficult-to-learn tool for both assessment and teaching.

In the next section, I describe students' last reflection, in which they described the most important lessons they learned from the course as a whole. They discussed young children's mathematical competence, the teaching of math to young children, the need to understand children's thinking, and the centrality of the clinical interview method.

The final reflections: lessons learned about early math education

At the end of the course, I posed these questions to the students:

> Based on what you have learned, what is the most important thing you would want teachers to know about teaching early mathematics?

> How has the course changed your views about early math education? How has the course changed your approach to teaching math to young children?

I begin with an account of the student reflections on what they learned about early math education and children's capabilities and after that turn to their reflections on the course itself.

Little children can learn math

Students were clearly convinced that young children are capable of learning much more math than is ordinarily supposed.

> The most important thing I learned in this class is the importance of nurturing mathematics in the young mind. Children can do more than many of us give them credit for and if we are able to foster their innate interests in numbers and math, we can really set them up for success and a better attitude toward the subject when they grow older.

> **

> In the beginning of the semester, I was wondering how little kids learn big math. Now that I am very sure little kids CAN learn big math!

The first lesson is about children's competence: they already have an everyday math of some power (alongside some weaknesses) and can learn a lot more. It's not true that abstract (but everyday) math is beyond children's capabilities.

Teaching math to little children is necessary and appropriate

Students agreed that math teaching—along with free play—is necessary for pre-school children.

> The most important thing that I would want teachers to know about teaching early mathematics is that it should and can be incorporated with this age group. Teachers need to know that children are capable of learning those skills and that teaching it can be developmentally appropriate.

> **

> I always thought that early math education only focused on learning how to recognize numbers and count. I did not give enough credit to children and their abilities to solve math problems ..., I never really thought of the importance of teaching it to really young children.

> **

> Early childhood teachers focus more on literacy than math, I'll make sure math is focused on just as much as literacy.

The second lesson is that teaching can be "developmentally appropriate," as the first student wrote. Of course, this teaching needs to be of high quality and should not focus predominantly on written or dull math tasks (like memorizing number facts). Of course, children learn from play too. But yes, we need intentional teaching of math, indeed perhaps as much as of literacy.

Understanding the child's thinking is key to successful teaching

Many students came to understand that children's thinking may be different from ours and that gaining insight into it is a prerequisite for teaching.

> I learned a lot about differences between what a child says, shows, knows, and means .... The most important thing I want teachers to know about teaching early mathematics is that it's our job to understand the child's thinking behind their answer. When we do understand their thinking, then we are beginning to understand how they are approaching mathematics and then we can truly TEACH!

The next student makes the point that it's important to understand the thinking of the child who gets a correct answer (as well as the child who gets a wrong answer).

> I think I've learned that one of the most important things for a teacher to be aware of is that although a child is able to correctly solve a problem, it may not necessarily mean that the child has a complete understanding of the underlying mathematical concepts of that problem.

This student raises the issue of group norms and the individual.

> I think the most important thing for teachers to know is that you really need to get to know each individual child and what they understand and do not understand about math. Broad ideas about what is appropriate for an age group are all well and good, but to truly teach a kid you need to understand that kid.

In brief, the third lesson is the basic principle of cognitive psychology (from Piaget onward): it is crucial to dig beneath surface behavior (like the right or wrong answer) to uncover the individual child's thinking. Norms and developmental trajectories are useful statistical ideas but do not necessarily illuminate the individual's thinking.

The clinical interview method is central

As the earlier reflections showed, the students were virtually unanimous in their appreciation of the clinical interview. They reiterated this point in their final reflections.

To me, the biggest takeaway ... is that there is a wealth of knowledge to be gained from interviewing children and interpreting their mathematical thinking.

**

I believe I will use the clinical interviews as an amazing tool to assess and inform my own teaching of math with young children. ... I will value all my students process of getting to an answer more than the simple result. Asking questions as "how did you know?" "how did you do it?" "can you teach me to do it?" will provide students with analysis of their own understanding and additionally will give students an awareness of their own thinking.

One student put her finger on the heart of the matter:

The most important thing I learned from the class is I am able to learn about how children think mathematics [sic].

In brief, the fourth lesson is that the clinical interview is an essential tool that can provide teachers with insight into how children think and what they have learned. How can you teach without understanding children's minds?

To summarize, at the end of the course, students reflected on broad lessons learned: young children are capable of learning more math than ordinarily supposed, deliberate teaching is developmentally appropriate, successful teaching requires an understanding of children's thinking, and the clinical interview can provide insight into children's thinking. I believe that these general lessons were in the nature of core beliefs, long remembered, that students will use to guide their teaching.

In the next section, I describe what students learned about the course itself and their suggestions about how to improve it. They wrote about the importance of videos, the value of conducting clinical interviews, the importance of the weekly reflections, and the need for a stronger emphasis on teaching.

Reflections on the course pedagogy

We turn to student views of the pedagogy of the course and needed improvements.

Analysis of videos can be transformative

The heart and soul of my pedagogy is to have students analyze videos intended to exhibit children's thinking in a clinical interview or naturalistic situation (Ginsburg, 2017). Here is what some students wrote about this approach.

I enjoyed the opportunities in class to analyze videos of clinical interviews with children. I believe that was the most vital part of the course, it really helped me understand how to better understand and consider young children's thinking.

**

I think that the most valuable things were the videos we watched. It was nice to see what we were reading play out in "real life."

**

Our readings were very important, yet seeing children interact with mathematical concepts in a video helped me understand how to be more intentional in my own teaching of mathematical concepts.

In brief, analyzing videos of children's thinking provides students with an exciting adventure in linking theory and practice. Students can see thinking in action; can see vivid examples of what is discussed in the research papers; and can begin to consider how teaching can relate to, be based on, and foster the thinking they see.

Conducting an interview is enormously valuable

For both the mid-term and final papers, students were required to conduct a clinical interview with an individual child.

To be honest even though it was a lot of work; I found the midterm most valuable. Making the videos, and writing the paper helped me to look at children's answers more than just correct or incorrect, critically.

**

Lastly, I'm glad I had the opportunity to conduct a clinical interview myself for the midterm project, so I was able to experience the mindset and apply methods that are necessary to gain more insight into the minds of children.

**

The midterm clinical interview assignment really helped put everything we were learning and talking about into perspective.

In brief, conducting an interview gave students the opportunity to apply what they had read (the academic papers) and seen (the class videos) to the task of uncovering the thinking underlying a very live child's performance. This was a melding of the academic and the practical, the ivory tower and the world of children, a synthesis of theory and practice.

The reflections helped students extend learning

Several students wrote that the reflection assignments were important stimulants to a deeper examination of the topic under consideration.

> The weekly reflections were very helpful for me. As a quiet participant in class, I felt like this was a good platform for me to extend what I learned in class and delve deeper into how I felt about what we were learning & discussing.

The need for a stronger focus on teaching

Some students reflected on how the course could place a stronger emphasis on teaching.

> If there was something I could request, because this course is a requirement for the Early Childhood Education program, having lessons that are focused on teaching math specifically to the early childhood children would have made the course more useful and relatable.

> **

> There were many aspects of this course which I felt have been helpful, however I believe that there truly needs to be more of an emphasis on how to teach math lessons. For the early childhood students this is the only math related course we take, and although this was great to learn about development, at least one day devoted to lesson implementation would have been greatly appreciated.

My perhaps defensive response to this is that I could not cover everything in one course. Over the years, I incorporated more and more about curriculum and teaching, but probably I could have done even more. In particular, I could have devoted more attention to critical examination of various curricula currently in use.

Other students reflected on their entire program.

> For ECE students, our mandatory classes are pretty much lined up for us (we only get one elective choice class) and they all tend to fall in line with pedagogical theory and ELA [English Language Arts] importance (writing, reading, literacy skills). It was really great for me to finally take a class that discussed something new, and equally as important! For my entire first year [in the program], it seemed like everyone forgot that teachers have to teach kids math as well!

> **

> I think that early childhood majors (like myself) should be required to take more than one class. Quite frankly, I do not think that one class is nearly enough to prepare you for kindergarten through second grade math instruction.

This final set of comments attests to a reality of higher education: the system offers insufficient training in teaching mathematics to young children. At least in the United States, education courses are skewed toward literacy, a topic with which prospective teachers are comfortable and familiar, rather than toward math, the topic with which they need the most help because they often fear it and know next to nothing about it (Hyson & Woods 2014).

Suggestions

I know that my course was distinctive. I had the luxury of focusing the entire course on "The Development of Mathematical Thinking." Many instructors may not want to do this even if they could. My students may not be like those in other programs. My approach to teaching prospective teachers may not be similar to yours.

How then can my experience with this course, and my students' reflections on it, inform your approach to the teaching of early childhood math education? Here are some suggestions, most of which draw upon my students' reflections on the course pedagogy.

Analysis of children's thinking

My students were captivated by videos illustrating children's thinking. Most videos showed clinical interviews; other videos showed everyday behavior. In both cases, I helped the students examine the videos closely and develop sound interpretations of children's thinking. I believe that this kind of careful analysis and interpretation should be a core element of the instructor's pedagogy (Ginsburg, 2017). Indeed, watching and analyzing carefully selected videos can be a magical experience for students and instructor alike. The observer/interpreter is a detective in an exciting mystery involving the child mind. I have met students who years later remember some of the videos they saw and analyzed in my classes. So I suggest that your teaching should include analyses of videos showing interviews and everyday behavior.

Clinical interviews

In addition to helping my students to analyze clinical interviews that I or others conducted, I required students to conduct, record, analyze, and interpret their own interviews both in mid-term and final papers. Going through this entire process can be a valuable and sometimes transformative experience, as some student reflections reveal. Doing the interview helped students see how their readings about research and theory were useful for interpreting the child's responses to the interviewer's probes. In other words, the interview helped students connect theory, readings, and their own practice. Learning to interview can be difficult and scary but can promote a skill that can well serve the classroom teacher (as well as the researcher). I suggest that you incorporate into your class at least some work on interviewing.

Reflections

Requiring students to reflect on their own learning, without fear of public exposure or grading, was much more valuable for my teaching than I had thought. Most importantly, the reflections gave me insight into what the students did and did not learn in the last class session and also gave me the opportunity to discuss their concerns at the outset of the next class session. In response to individual reflections, I could discuss such issues as how math fears were common, how students may have misinterpreted "math is all around us," and how play is not sufficient for math education. In this way, I was able to integrate my students' thinking into class discussion.

Students seemed to enjoy the process. They seemed quite happy, trying to hide their grins when I, trying to avoid looking at them, discussed their (anonymous) reflections. I suggest that you have students reflect on the previous class as often as you can. Reading the reflections and responding to them is well worth the effort.

Stories

Storybooks can play a very useful part in early childhood education, for teacher and child alike. As I mentioned, good math storybooks can be fun at the same time as they introduce mathematical ideas. Storybooks are not threatening to teachers averse to math education.

I made available appropriate storybooks for students to read when we discussed children's understanding of adding, shape, pattern, and the like. We discussed the concepts in the books and how to use them effectively in the classroom. This was a form of "instruction" my students could easily understand and appreciate. I think your students will feel the same way.

Focus on teaching

I spent more time on the analysis of thinking than on promoting it by deliberate instruction. Why? Because that was my personal preference and skill, as a result of my training as a developmental psychologist rather than as a math educator. My students felt that I should have devoted more time to the discussion of teaching. Well, it's too late now, but you can certainly focus both on teaching and children's thinking. You can examine activities from current curricula. You can have students analyze classroom videos of math "lessons" to shed light on the teacher's teaching strategy and its effect on the student's understanding. In my class, the final assignment was for students to interview an individual child about their understanding of some math concept, then teach the concept, and finally interview the child again to determine whether and what they may have learned. You can do something like this as well.

The system

As some students noted, their program provided the least help (in the form of courses) on the subject—math education—for which they required the most help. This is not uncommon (Hyson & Woods, 2014) and should be changed: there should be more math education courses for early childhood students.

But I would go further. As I taught my course, I had a fantasy that a talented early childhood teacher served as a co-instructor, and that my students visited the teacher's class on a regular basis. The teacher would comment on how the readings, the videos, and the lectures related to what she observed daily in her classroom and informed, or failed to inform, her teaching. She could discuss what the students saw in her classroom, and I could respond to that. The students could carry out their interviews with her children and could devise and implement lessons in her classroom. She could contribute to our class discussions about pedagogy and curriculum. This imaginary teacher-participant would be extremely highly appreciated! Is there a way to build her into our system of higher education?

A resource

Here is a resource that you can use in your early education courses. DREME (Development and Research in Early Math Education) is an organization, of which I am a member, that is devoted to creating materials useful for your work. The DREME website (https://dreme.stanford.edu) offers free videos, papers, storybook reading guides, analyses of children's thinking, examples of effective teaching, and much more. Check it out!

References

Blömeke, S., Gustafsson, J.-E., & Shavelson, R. J. (2015). Beyond dichotomies: Competence viewed as a continuum. *Zeitschrift für Psychologie, 223*(1), 3–13. https://doi.org/10.1027/2151-2604/a000194

Carpenter, T. P., Franke, M. L., Johnson, N. C., Turrou, A. C., & Wager, A. A. (2017). *Young children's mathematics: Cognitively guided instruction in early childhood education.* Portsmouth, NH: Heinemann.

Clements, D. H., & Sarama, J. (2014). *Learning and teaching early math: The learning trajectories approach* (2nd ed.). New York, NY: Routledge.

Ginsburg, H. P. (1989). *Children's arithmetic* (2nd ed.). Austin, TX: Pro-Ed.

Ginsburg, H. P. (1997). *Entering the child's mind: The clinical interview in psychological research and practice.* New York, NY: Cambridge University Press.

Ginsburg, H. P. (2016). Anna counts. http://prek-math-te.stanford.edu/counting/anna-counts (accessed 02/02/2021).

Ginsburg, H. P. (2017). Pedagogy of the video clip. http://prek-math-te.stanford.edu/overview/pedagogy-video-clip (accessed 02/02/2021).

Ginsburg, H. P., Inoue, N., & Seo, K.-H. (1999). Preschoolers doing mathematics: Observations of everyday activities. In J. Copley (Ed.), *Mathematics in the early years* (pp. 88–99). Reston, VA: National Council of Teachers of Mathematics.

Ginsburg, H. P., Uscianowski, C., Carrazza, C., & Levine, S. C. (2020). Print and digital picture books in the service of young children's mathematics learning. In O. N. Saracho (Ed.), *Handbook of research on the education of young children* (4th ed., pp. 85–98). New York, NY: Routledge.

Herts, J. B., Beilock, S. L., & Levine, S. C. (2019). The role of parents' and teachers' math anxiety in children's math learning and attitudes. In I. C. Mammarella, S. Caviola, & A. Dowker (Eds.), *Mathematics anxiety: What is known and what is still to be understood* (pp. 190–210). Oxon: Routledge.

Hirsch, E. S. (1996). *The block book.* Washington, DC: National Association for the Education of Young Children.

Hyson, M., & Woods, T. A. (2014). Practices, knowledge, and beliefs about professional development. In Ginsburg, H. P., Hyson, M. and Woods, T. A. (Eds.), *Preparing early childhood educators to teach math* (pp. 29–52). Baltimore, MD: Paul H. Brooks Publishing Co.

Kontos, S. (1999). Preschool teachers' talk, roles, and activity settings during free play. *Early Childhood Research Quarterly, 14*(3), 363–382.

Lozada, M., & Carro, N. (2016). Embodied action improves cognition in children: Evidence from a study based on Piagetian conservation tasks. *Frontiers in Psychology, 7*, 393. https://doi.org/10.3389/fpsyg.2016.00393

Singer, D. G., Golinkoff, R. M., & Hirsh-Pasek, K. (Eds.) (2006). *Play = Learning: How play motivates and Enhances children's cognitive and social-emotional growth.* New York, NY: Oxford University Press.

Chapter 2

Preschool teachers' mathematical pedagogical content knowledge and self-reported classroom activities

Joke Torbeyns, Febe Demedts, Fien Depaepe

Introduction

Children's early mathematical development is pivotal for their later academic performances and professional career (Duncan et al., 2007). Cumulative evidence indicates that the mathematical classroom activities provided in preschool are important for the acquisition of early and later mathematical competencies (Claessens, Engel, & Curran, 2014; Engel, Claessens, Watts, & Farkas, 2016; Klibanoff, Levine, Huttenlocher, Vasilyeva, & Hedges, 2006; Lehrl, Kluczniok, & Rossbach, 2016; Melhuish et al., 2013; Piasta, Logan, Pelatti, Capps, & Petrill, 2015). Unfortunately, preschool teachers infrequently offer classroom activities to support young children's mathematical development, although differences between preschool teachers are observed (Claessens et al., 2014; Engel et al., 2016; Klibanoff et al., 2006; Piasta, Pelatti, & Miller, 2014). One of the potential explanations for the generally low but inter-individually different organisation of classroom activities in the domain of early mathematics refers to preschool teachers' professional competence, and specifically potential weaknesses in preschool teachers' knowledge and skills to teach early mathematics to young children (Li, 2020; Piasta et al., 2015). The present study aimed to empirically address this hypothesis by (a) investigating the acquisition of pedagogical content knowledge (PCK) in the domain of mathematics in pre-service and in-service preschool teachers and (b) analysing the association between in-service preschool teachers' PCK and the classroom activities they report to provide to their pre-schoolers.

Theoretical and empirical framework

Teachers' professional competence

Cumulative evidence indicates that teachers' professional competence is pivotal for instructional quality and student learning (Burroughs et al., 2019; Hattie, 2009). Teachers' professional competence is a multi-dimensional construct, consisting of domain-specific and domain-general knowledge, skills and beliefs that

DOI: 10.4324/9781003172529-2

are foundational for observable teaching behaviour and instructional quality (Baumert & Kunter, 2013; Blömeke, Gustafsson, & Shavelson, 2015). Blömeke and colleagues (2015) define teachers' professional competence along a continuum consisting of three dimensions: from dispositions via situation-specific skills to observable teaching behaviour. The first dimension involves teachers' *dispositions*, referring to both cognitive and affective-motivational facets (i.e. professional knowledge and beliefs). Teachers' cognitive facets are defined as domain-specific and domain-general knowledge, including domain-specific subject-matter knowledge (content knowledge or CK), domain-specific pedagogical content knowledge (PCK; knowledge of students' [mis]conceptions and of effective instructional strategies in the subject-matter domain) and domain-general pedagogical knowledge (GPK; e.g. knowledge of general strategies for classroom management). Teachers' affective-motivational facets entail their domain-specific and domain-general professional beliefs, including teachers' ideas about the nature of the subject-matter domain, about learning and teaching in this domain and about learning and teaching in general. The second dimension of the model refers to teachers' *situation-specific skills*, i.e. the cognitive processes before, during and after actual classroom behaviour. Teachers notice events that happen in the classroom (perception), make sense of these events (interpretation) and decide on how to respond (decision making). Both teachers' dispositions and situation-specific skills contribute to the third dimension of the model: teachers' *actual classroom behaviour*. As argued by Kunter and colleagues (2013), teachers' classroom behaviour mediates the frequently observed association between, on the one hand, their professional knowledge and beliefs and, on the other hand, students' learning outcomes.

Teachers' professional competence in the domain of mathematics has been studied intensively at the *elementary and secondary school* level (e.g. Baumert & Kunter, 2013; Baumert et al., 2010; Kaiser et al., 2017; Kersting, Givving, Thompson, Santagata, & Stigler, 2012; Krauss et al., 2020; Kunter et al., 2013; Yang, Kaiser, König, & Blömeke, 2020). Together, these studies provide empirical support for the structure of the above model. Teachers' professional knowledge and professional beliefs were shown to be distinct but related facets of teachers' professional competence, both contributing to their situation-specific skills and teaching behaviour (e.g. Baumert et al., 2010; Krauss et al., 2020; Yang et al., 2020). Importantly, teachers' PCK proved foundational for instructional quality and student achievement: Higher teacher PCK was positively associated with both effective instructional practices and stronger student learning outcomes (e.g. Baumert et al., 2010; Krauss et al., 2020; Kunter et al., 2013). Moreover, the association of instructional quality with teachers' PCK was shown to be larger than its association with teachers' CK (e.g. Baumert et al., 2010).

Given the pivotal role of PCK for effective instructional practices or classroom activities, the acquisition of this important type of knowledge during initial teacher training requires further research attention. Schmidt, Cogan, and

Houang (2011) identified four main types of opportunities to learn (OTL) professional knowledge in the domain of mathematics during teacher training, namely (a) theoretical courses on CK, (b) theoretical courses on PCK, (c) theoretical courses on GPK and (d) practical experiences. Theoretical courses on PCK in the domain of mathematics were identified as the major source for PCK development during initial teacher training at the elementary and secondary school level (Blömeke, Buchholtz, Suhl, & Kaiser, 2014; Kleickmann et al., 2013; Qian & Youngs, 2016).

Preschool teachers' professional competence in the domain of early mathematics only recently attracted the attention of researchers (cf. Dunekacke, Jenßen, Eilerts, & Blömeke, 2016; Torbeyns, Verbruggen, & Depaepe, 2020). Although limited in number, these first studies confirm the major findings of the studies in elementary and secondary school teachers. First, preschool teachers' professional knowledge and beliefs were shown distinct but related facets of their professional competence, and they both contributed to preschool teachers' situation-specific skills (Dunekacke et al., 2016; Oppermann, Anders, & Hachfeld, 2016; but see Anders & Rossbach, 2015). Second, individual differences in preschool teachers' PCK were positively related to their teaching behaviour and instructional quality, as indicated by the classroom activities preschool teachers offered to their pre-schoolers (J. Lee, Meadows, & Lee, 2003; McCray & Chen, 2012). Third, theoretical courses on PCK were shown to improve pre-service and in-service preschool teachers' acquisition of PCK (Blömeke, Jenßen, Grassmann, Dunekacke, & Wedekind, 2017; Bruns, Eichen, & Gasteiger, 2017; Polly et al., 2017) as well as in-service teachers' classroom activities in the domain of early mathematics (Polly et al., 2017). Additionally, for in-service preschool teachers, PCK was positively associated with teachers' number of years of teaching experience at preschool (J. Lee, 2010; J. E. Lee, 2017). Unfortunately, even experienced in-service preschool teachers were shown to have specific weaknesses and difficulties related to their PCK (J. E. Lee, 2017; Li, 2020).

Mathematical classroom activities

Notwithstanding the pivotal role of effective mathematical classroom activities in preschool for children's early and later mathematical development, empirical studies on the occurrence of these activities and on its association with preschool teachers' PCK are scarce. First, focusing on the occurrence of classroom activities in the domain of early mathematics in preschool, empirical studies indicated that preschool teachers spend a limited amount of time on mathematical talk and mathematical activities during their daily classroom activities (Björklund & Barendregt, 2016; Claessens et al., 2014; Engel et al., 2016; Klibanoff et al., 2006; Piasta et al., 2014). Additionally, during their mathematical activities, preschool teachers mainly focused on the domains of number and counting.

Second, the few studies that investigated the association between preschool teachers' PCK and the classroom activities they offer in the domain of early mathematics (Lee et al., 2003; McCray & Chen, 2012; Polly et al., 2017) observed positive correlations between preschool teachers' PCK and the mathematical activities they (report to) offer to their pre-schoolers. The latter findings are in line with the available evidence at the elementary and secondary school level, revealing positive associations between teachers' PCK and the quality of the classroom instruction they provide (e.g. Baumert et al., 2010; Kunter et al., 2013; Krauss et al., 2020).

The present study

Taking into account the pivotal role of teachers' PCK for effective classroom activities and the limited number of studies on preschool teachers' PCK as well as its association with classroom activities, we aimed at (a) investigating the acquisition of PCK in a large sample of pre-service and in-service preschool teachers and (b) analysing its association with self-reported classroom activities. We included pre-service preschool teachers from the first to the final year of initial teacher training with different amounts of theoretical courses on PCK in the domain of early mathematics and practical internships. In-service preschool teachers all had fulfilled their theoretical training and practical internship periods related to the initial teacher training but differed in the number of years of teaching experience in preschool. As such, we were able to analyse the acquisition of PCK in preschool teachers differing in theoretical and practical baggage.

We conducted the study in Flanders, i.e. the Dutch-speaking part of Belgium. In Flanders, preschool is organised for children aged 2.5 until 6 years. Although preschool is not compulsory, about 97% of the Flemish children attend preschool (www.ond.vlaanderen.be). Preschool teachers are expected to stimulate the development of children's competencies in all curricular domains, including mathematics, in view of the developmental goals formulated by the Flemish government. Flemish preschool education typically involves informal learning situations, with age-appropriate play-based learning activities that focus on core competencies from different curricular domains in an integrated way.

Accordingly, Flemish preschool teachers are trained as generalists in a professional bachelor training programme at non-university teacher training institutes. The bachelor programme involves three years of training and accounts for 180 European Credit Transfer System (ECTS) credits (i.e. 60 ECTS credits per year of training), including 45 ECTS credits for practical internships. Pre-service preschool teachers have to follow theoretical courses in all curricular domains, including early mathematics, and in general pedagogy and psychology. Student internships typically focus on teaching competencies in all curricular domains (and not only mathematics), aiming for the informal and play-based stimulation of different core competencies in an integrated way.

Method

Participants

The sample consisted of 162 pre-service preschool teachers and 75 in-service preschool teachers (total sample of 237 participants). All participants gave active informed consent to participate in the study. Table 2.1 presents the total number and the number of female participants in the pre-service and in-service sample, as well as their age, educational track in secondary education and their year of training, respectively, number of years of teaching experience in preschool.

We recruited first- to third-year pre-service preschool teachers from two Flemish teacher training institutes. At the start of the study, 202 pre-service preschool teachers agreed to participate. As there is a strong selection between the first and second year of teacher training, we excluded 40 first-year students that did not succeed at the end of the academic year to prevent selection effects. The 162 remaining pre-service preschool teachers were 60 first-year, 65 second-year and 37 third-year students (see Table 2.1). We additionally recruited 75 in-service preschool teachers with at least one year of teaching experience in Flemish preschool. The in-service preschool teachers were coming from 23 different preschools, and they varied in their teaching experience from 2 until 35 years. Only ten in-service preschool teachers had attended at least one professional development course (lasting for a half or at most one day) in the domain of early mathematics during the past three years, with frequencies ranging from one up to five courses.

Although the pre-service preschool teachers followed their initial teacher training in two different institutes, the timing and the number of ECTS credits for PCK courses in the domain of early mathematics and for practical internships

Table 2.1 Number, age, educational track in secondary education, year of training (pre-service) and years of teaching experience (in-service preschool teachers)

Group	Year of training/teaching experience	N All	N Female	Age in years (SD)	Educational track secondary education AC	VO	AR	TE
Pre-service	1	60	58	20.12 (2.62)	8	19	2	31
	2	65	62	21.04 (1.92)	10	16	5	34
	3	37	34	24.47 (6.67)	8	12	3	14
	All	162	154	21.75 (4.34)	26	47	10	79
In-service	2–10	25	25	28.00 (3.70)	6	1	0	18
	11–25	24	24	38.53 (4.35)	7	2	0	15
	>25	26	26	50.27 (3.74)	13	0	0	13
	All	75	75	40.19 (10.52)	26	3	0	46

AC = Academic, AR = arts, TE = technical, VO = vocational.

Table 2.2 Timing and ECTS credits for theoretical PCK courses and practical internships per teacher training institute

Teacher training institute	Year of training	Theoretical courses on PCK in early mathematics			Practical internships		
		Semester 1	Semester 2	ECTS	Semester 1	Semester 2	ECTS
1	1	–	X	3	X	X	9
	2	X	–	3	X	X	15
	3	–	–	0	X	X	21
2	1	–	X	3	X	X	6
	2	X	–	3	X	X	14
	3	–	–	0	X	X	25

did not differ between the institutes. We therefore grouped the pre-service preschool teachers from the different teacher training institutes in all our analyses. Table 2.2 reflects the concrete timing and the exact number of ECTS credits for theoretical courses on early mathematics PCK and for practical internships per teacher training institute. It is important to note here that all theoretical courses on PCK in the domain of mathematics were offered in the second semester of the first year of training and in the first semester of the second year of training. Consequently, the theoretical baggage related to PCK in the domain of mathematics was highly similar for the second-year and third-year pre-service preschool teachers at the time of the data collection (i.e. at the start of the second semester). As the third-year students did not yet start their nine-week internship period in the last semester of their teacher training, second-year and third-year students did not only have similar theoretical baggage but also highly analogous practical experiences. We therefore included second- and third-year students as one group of students with similar theoretical and practical baggage in our analyses.

We used the scenario-based instrument of Torbeyns and colleagues (2020) to investigate preschool teachers' *PCK* in the domain of mathematics. This instrument was developed on the basis of the scenario-based instrument of Gasteiger, Bruns, Benz, Brunner, and Sprenger (2020), with adaptations in view of the Flemish preschool curriculum for early mathematics. The instrument consists of five scenarios or descriptions of mathematically rich preschool situations addressing the three major domains of the early mathematics preschool curriculum in Flanders: two scenarios in the domain of number (number, counting), two scenarios in the domain of measurement (measurement of volume, measurement of time) and one scenario in the domain of geometry (patterns). Each scenario is followed by a series of ten multiple-choice items addressing preschool teachers' knowledge of pre-schoolers' mathematical competencies and of effective teaching strategies in the situation. The scenario-based instrument thus consists of 50 items that are scored dichotomously (correct, incorrect), resulting in a maximum score of 50. The questions related to preschool teachers' knowledge of pre-schoolers' mathematical competencies have to be answered by indicating whether (1) the pre-schooler masters the competency in the specific situation, (2) the pre-schooler does not master the competency in the specific situation, (3) the pre-schoolers' mastery of the competency cannot be observed in the specific situation, or (4) the preschool teacher does not know the answer to the question. The questions related to effective teaching strategies require the pre-school teacher to select the most appropriate strategy to stimulate pre-schoolers' mathematical competencies in the specific situation from a series of four options, and the additional option 'I don't know'. On the basis of test- and item-reliability analyses, we excluded four items from the instrument (i.e. one item related to the domain of number and three items related to the domain of measurement). The final instrument thus consisted of 46 items, with a maximum score of 46. The Cronbach's alpha value for the instrument as a whole was .77, but the Cronbach's alpha values for the five scenarios separately did not reach the required level of .70 (values ranging from .15 to .68).

Additionally, we collected information on the *classroom activities in the domain of early mathematics* that the in-service preschool teachers offered via an adapted version of the questionnaire of Björklund and Barendregt (2016). The adapted questionnaire only addresses preschool teachers' mathematical space, with focus on the specific content domains addressed in the PCK instrument (i.e. domain of number: numbers, counting; domain of measurement: measurement of volume, measurement of time; domain of geometry: patterns). For each of these five specific content domains (numbers, counting; measurement of volume, measurement of time; patterns), the classroom activities that the preschool teachers engage in are questioned using six items, namely preschool teachers' (a) provision of specific mathematical content in the classroom, (b) attention for specific mathematical content in daily classroom activities, (c) stimulation of pre-schoolers to attend to specific mathematical content in daily classroom activities, (d) observation of pre-schoolers' tendencies to explore and use mathematics, (e) support to enable pre-schoolers to communicate about specific mathematical content and (f) support to enable pre-schoolers to problematise specific mathematical content. For each item, the frequency of occurrence is indicated on a 5-point Likert scale, ranging from never to daily (respectively, 0–4). The maximum score on the questionnaire is (5 [domains] × 6 [items per domain] × 4 [Likert scale 0–4]) 120. Cronbach's alpha for the questionnaire was .94. The Cronbach's alpha values for the five subdomains were sufficient, with values ranging from .75 to .89.

Pre-service and in-service pre-schoolers' PCK was assessed at the start of the second semester (i.e. January–March). The data collection in the pre-service pre-school teachers was organised collectively at their teacher training institutes. For the in-service preschool teachers, the data collection was done individually or in small groups and at their preschools. Both the PCK instrument and the questionnaire on classroom activities were offered in paper-and-pencil format. As the participating pre-service preschool teachers had only engaged in a limited amount of internship teaching activities, following the guidelines of their supervisors at the teacher training institute and in-service preschool teachers, we did not offer them the questionnaire on classroom activities.

Analyses

To answer the first major aim of the present study, we analysed pre-service and in-service preschool teachers' PCK via Kruskal–Wallis analyses and Spearman's rho correlation analyses, because the assumption of normality for parametric analysis techniques was violated. Given the low reliabilities of the five scenarios separately, we focused on the general level of PCK as indicated by the total test score. We first compared the level of PCK among three groups of preschool teachers: (a) first-year pre-service preschool teachers who just started their first theoretical course on PCK in the domain of early mathematics and had not yet engaged in active internships, (b) second-year and third-year pre-service preschool teachers who had fulfilled all theoretical courses on PCK in the domain of early mathematics and had limited practical experiences and (c) in-service preschool teachers with at least one year of teaching experience in preschool. We next computed the association between in-service preschool teachers' PCK and their number of years of teaching experience in preschool.

In view of the second major aim, we computed the association between in-service preschool teachers' PCK and the classroom activities they reported to offer via Spearman's rho correlations, because the assumption of normality for a parametric Pearson's correlation was violated. We analysed their association for the classroom activities in general (all five subdomains) and for the five subdomains separately. We additionally explored the association between the reported classroom activities and in-service preschool teachers' number of years of teaching experience in preschool using Spearman's rho correlations. As mentioned above, we did not collect information about the pre-service preschool teachers' classroom activities, as they had only limited internship teaching experiences.

Results

Preschool teachers' PCK

We observed large individual differences in preschool teachers' PCK, with scores ranging from 11 to 44 on a maximum of 46. First-year pre-service preschool teachers, second- and third-year pre-service preschool teachers and in-service preschool teachers significantly differed in their scores on the PCK instrument, $H(2) = 57.11$, $p < .001$. Pairwise group comparisons using Mann–Whitney tests with Bonferroni correction and a significance level of .0167 (Field, 2018) revealed that first-year students scored lower on the PCK instrument than both second- and third-year students, $U = 1535.00$, $r = -.42$, $p < .001$, and in-service preschool teachers, $U = 600.50$, $r = -.63$, $p < .001$ (respectively, $M_{\text{pre-1}} = 28.47$ [SD = 5.20], $M_{\text{pre-2-3}} = 33.30$ [SD = 5.17] and $M_{\text{in}} = 35.61$ [SD = 4.95]). Next, second- and third-year students obtained significantly lower scores than in-service preschool teachers, $U = 2773.00$, $r = -.24$, $p = .002$. Taken together, first-year pre-service preschool teachers had lowest PCK in the domain of early mathematics and in-service preschool teachers had highest, albeit not yet perfectly mastered, PCK in this domain.

We additionally evaluated the association between in-service preschool teachers' PCK and their years of preschool teaching experience via Spearman's rho correlation analysis. This additional analysis revealed no significant correlation between these two variables, rho = .001, $p = .50$ (one-tailed).

Mathematical classroom activities

In-service preschool teachers differed in the amount of classroom activities they reported to engage in in the domain of early mathematics, as indicated by a range in score from 33 to 113 (on a maximum of 120). Repeated measurements ANOVA (with Greenhouse–Geisser correction) revealed differences between the five domains, $F(3.29, 233.57) = 96,83$, p < .001. Post-hoc comparisons with Bonferroni correction indicated that in-service preschool teachers reported most activities related to number ($M = 20.23$, SD = 4.54, indicating weekly to daily occurrence of activities in this mathematical domain), followed by counting ($M = 16.69$, SD = 4.55, i.e. monthly to weekly) and measurement of time ($M = 16.40$, SD = 5.20); they engaged least in classroom activities related to measurement of volume ($M = 12.41$, SD = 4.57, reflecting on occurrence of max. once a month) and patterns ($M = 11.31$, SD = 4.38). We observed positive associations between teachers' classroom activities in the five different domains, ranging from .31 to .75 (all $ps < .01$). This indicates that in-service preschool teachers who frequently engaged in mathematical activities in one domain also did so in the other domains.

Next, the association between in-service preschool teachers' PCK and the mathematical classroom activities they reported was significant, rho = .23, $p = .03$ (one-tailed), which can be considered as a medium effect (Cohen, 1988). In-service preschool teachers with higher levels of PCK thus reported more mathematical activities in their classrooms than their colleagues with lower levels of PCK. These associations were also observed for the domains of number (rho = .18, $p = .06$), counting (rho = .22, $p = .03$) and measurement of time (rho = .29, $p < .01$), but not for the domains of measurement of volume (rho = .15, $p = .11$) and patterns (rho = .07, $p = .28$).

Finally, as was the case for PCK, we did not find an association between the occurrence of mathematical activities in the classroom and teachers' years of teaching experience, rho = .16, $p = .08$ (one-tailed). So, teachers with more years of teaching experience did not offer more mathematical classroom activities than their colleagues with less years of teaching experience.

Discussion

The present study aimed to complement current understanding of preschool teachers' competence, and especially their PCK, in the domain of early mathematics and of the association of the former with the classroom activities they provide in this domain. We therefore analysed the level of PCK in a large sample of pre-service and in-service preschool teachers who differed in the amount of theoretical courses on early mathematics PCK as well as the practical teaching experiences they had fulfilled, namely (a) pre-service preschool teachers who just had started their theoretical and practical training and had received hardly any theoretical and practical opportunities to acquire the relevant knowledge and skill, (b) pre-service preschool teachers who had completed all theoretical courses on mathematical PCK but had limited practical experiences and (c) in-service preschool teachers who completed all theoretical training and who had at least one year of practical teaching experience in preschool. We additionally investigated the classroom activities that the latter group of preschool teachers reported to engage in. Our findings contribute to current insights into the acquisition of PCK in preschool teachers and its association with daily classroom activities in the domain of early mathematics. Moreover, they raise questions for future studies on this topic and suggestions for current initial teacher training and professional development initiatives.

PCK in the domain of early mathematics

The first major finding of the present study relates to the acquisition of PCK in preschool teachers. Pre-service preschool teachers who just started their initial teacher training had limited PCK in the domain of early mathematics. During initial teacher training, the level of PCK increased, as evidenced by the higher PCK scores of the second-year and third-year pre-service preschool teachers. The latter group of students had fulfilled all theoretical courses related to PCK in the domain of early mathematics and had limited practical teaching experience in preschool. As such, these findings point to the pivotal role of theoretical courses on early mathematics PCK for the acquisition of this important type of knowledge, confirming previous findings at the elementary and secondary school level (Blömeke et al., 2014; Kleickmann et al., 2013; Qian & Youngs, 2016) and recent findings at the preschool level (Blömeke et al., 2017).

Our findings related to the contribution of practical teaching experience to the acquisition of PCK are not unequivocal: The higher level of PCK in in-service preschool teachers with at least one year of teaching experience compared to the sample of pre-service preschool teachers suggests that practical teaching experience contributes to the further development of PCK at the end of and after initial teacher training. However, the increase in the level of PCK did not result in an almost perfect mastery of this important type of knowledge (about three-fourth of the items correctly answered, with inter-individual differences in the acquired level of PCK). Moreover, more years of teaching experience did not result in similar increases in the level of PCK in this group of in-service preschool teachers after initial teacher training. The weaknesses in in-service preschool teachers' PCK in the domain of early mathematics were also observed in previous studies on that topic (J. Lee, 2010; J. E. Lee, 2017; Li, 2020; McCray & Chen, 2012). However, the absence of an association between PCK and years of teaching experience is not in line with previous findings at the preschool level (J. Lee, 2010; J. E. Lee, 2017).

One possible explanation for the latter finding relates to in-service preschool teachers' *deliberate practice* (following the notion of the 'reflective practitioner'; Schön, 1983) as basis for further competence development (König, Blömeke, & Kaiser, 2015). As outlined above, the participating third-year pre-service preschool teachers had not yet started their nine-week internship period in the final semester of their initial teacher training, whereas in-service preschool teachers had fulfilled this internship period and had at least one year of additional teaching experience in preschool classrooms. As the nine-week internship period involves the integration of previous theoretical courses and concrete classroom practices, supervised by both a lecturer of the initial teacher training institute and the classroom teacher of the classroom the internship is conducted in, it can be considered as practical teaching experience complemented with theoretical reflection in and upon this experience (cf. the notions of 'reflection-in-action' and 'reflection-on-action'; Schön, 1983). After the finalisation of their initial teacher training, these theoretical reflections are not systematically organised for in-service preschool teachers, with the exception of some short professional development courses. The participating in-service preschool teachers indicated that they hardly followed any professional development course in the domain of early mathematics, limiting the possibilities to integrate critical theoretical reflection into actual teaching practices. Taking into account the limited participation to professional development courses in the domain of early mathematics in the current sample and the positive findings of continuous professional development focusing on theoretical courses in the domain of early mathematics (Bruns et al., 2017; Polly et al., 2017), it seems beneficial to organise compulsory continuous development programs for preschool teachers in the domain of early mathematics, with particular attention for preschool teachers' PCK in the domain of early mathematics (see also Li, 2020; Li, McFadden, & DeBey, 2019). Although theoretical courses on early mathematics PCK were shown to enhance the acquisition of this important type of knowledge, the application of the theoretical content into real classroom settings and the critical reflection upon this integration are needed to realise effective changes in current early childhood mathematics education (Piasta et al., 2015; Thornton, Crim, & Hawkins, 2009). In other words, the inclusion of the core components of deliberate practice into compulsory continuous development programs can pave the way for the required improvements in preschool teachers' PCK in the domain of early mathematics.

It is important to note here that our findings on the acquisition of PCK during and after initial teacher training are based on cross-sectional data. As we compared the amount of acquired PCK between different cohorts, differing in their finalisation of major theoretical and practical OTL this important type of knowledge, we cannot draw strong conclusions on the development of PCK during teacher training and beyond. The actual development of pre-service and in-service preschool teachers' PCK requires longitudinal studies, with repeated measurements of PCK in the same cohort of – first – pre-service and – next – in-service preschool teachers. Such longitudinal designs allow to refine our understanding of the development of preschool teachers' PCK in association with the theoretical and practical OTL this knowledge offered during initial teacher training and beyond.

Classroom activities in the domain of early mathematics

The present study also resulted in new insights into the occurrence of classroom activities in the domain of early mathematics and its association with preschool teachers' PCK. As was done in previous studies in this domain (e.g. Björklund & Barendregt, 2016), we relied on teachers' self-reported practices and did not engage in (time-consuming) observations of actual classroom practices. Although the latter type of data might reveal more accurate descriptions of classroom activities, the association between teachers' self-reported and observed classroom practices was empirically confirmed in previous studies, also in the domain of preschool mathematics (Polly et al., 2017).

First, as was observed in previous studies (Björklund & Barendregt, 2016; Claessens et al., 2014; Engel et al., 2016; Klibanoff et al., 2006; Piasta et al., 2014), our findings indicated differences in the (reported) classroom activities in the domain of early mathematics between preschool teachers and between mathematical domains. When providing mathematical activities to their pre-schoolers, teachers preferred to focus on the domain of number, offering less opportunities to acquire core competencies in the domain of measurement and patterns as well (cf. Claessens et al., 2014; Engel et al., 2016). The occurrence of such activities was not associated with the years of teaching practice of the preschool teachers, but with their level of PCK. It is important to note here that the association between preschool teachers' PCK and mathematical classroom practices was observed for the 'basic' domains of number and counting, and to a lesser extent for the 'more complex' domains of measurement and patterns. The absence of an association in these more complex domains might be due to the generally low occurrence of mathematical activities in these domains and the related lower variability, making it methodologically difficult to observe an association. Additionally, it might be explained on the basis of specific weaknesses in preschool teachers' PCK in particularly these more complex mathematical domains. As the PCK instrument did not allow analyses for the different mathematical domains separately, the latter explanation needs to be addressed in future studies on this topic.

In sum, these findings again point to the importance of the development of a sufficiently high level of PCK during and after initial teacher training: Higher levels of PCK are associated with generally higher frequencies of classroom activities in the domain of early mathematics. As our study was correlational in nature, we cannot draw any causal conclusions about this association. However, the recent study of Polly and colleagues (2017) provides the first building blocks to further disentangle the direction and the causal nature of this association. As discussed in Polly and colleagues (2017), a carefully designed intervention programme focusing on preschool teachers' PCK cannot only enhance the latter type of knowledge, but also preschool teachers' reported and observed classroom activities, and, as such, pre-schoolers' mathematical competencies. Based on the available correlational evidence from also our study, and the positive findings of Polly and colleagues (2017), it is important to further unravel the direction and causal nature of the associations between preschool teachers' PCK and their classroom activities in the domain of early mathematics. In these studies, special attention should be paid to core questions related to the foundational content of such programs, the required practical experiences and the needed supervision to really change preschool teachers' knowledge and classroom activities in the domain of early mathematics. Moreover, the effectiveness of such programs for pre-schoolers' early mathematical development needs further study as well (but see the first positive findings of Polly et al., 2017; Tirosh, Tsamir, Levenson, & Tabach, 2011).

Conclusion

Taken together, our findings point to the role of both theoretical courses and 'reflective' practical experiences to the acquisition of preschool teachers' PCK. Moreover, our findings also support the assumed contribution of preschool teachers' PCK for providing mathematical activities at preschool. Importantly, as more years of teaching experience of in-service preschool teachers did not result in similar increases in PCK and mathematical activities at preschool, the present study highlights the need for continued compulsory professional development for in-service preschool teachers, including both theoretical and practical OTL.

References

Anders, Y., & Rossbach, H. G. (2015). Preschool teachers' sensitivity to mathematics in children's play: The influence of math-related school experiences, emotional attitudes, and pedagogical beliefs. *Journal of Research in Childhood Education*, *29*(3), 305–322. https://doi.org/10.1080/02568543.2015.1040564

Baumert, J., & Kunter, M. (2013). The COACTIV model of teachers' professional competence. In M. Kunter, J. Baumert, W. Blum, U. Klusmann, S. Krauss, & M. Neubrand (Eds.), *Cognitive activation in the mathematics classroom and professional competence of teachers* (pp. 25–48). https://doi.org/10.1007/978-1-4614-5149-5

Baumert, J., Kunter, M., Blum, W., Brunner, M., Voss, T., Jordan, A., Klusmann, U., Krauss, S., Neubrand, M., & Tsai, Y. M. (2010). Teachers' mathematical knowledge, cognitive activation in the classroom, and student progress. *American Educational Research Journal*, *47*(1), 133–180. https://doi.org/10.3102/0002831209345157

Björklund, C., & Barendregt, W. (2016). Teachers' pedagogical mathematical awareness in Swedish early childhood education. *Scandinavian Journal of Educational Research*, *60*(3), 359–377. https://doi.org/10.1080/00313831.2015.1066426

Blömeke, S., Buchholtz, N., Suhl, U., & Kaiser, G. (2014). Resolving the chicken-or-egg causality dilemma: The longitudinal interplay of teacher knowledge and teacher beliefs. *Teaching and Teacher Education*, *37*, 130–139. https://doi.org/10.1016/j.tate.2013.10.007

Blömeke, S., Gustafsson, J. E., & Shavelson, R. J. (2015). Beyond dichotomies: Competence viewed as a continuum. *Zeitschrift fur Psychologie/Journal of Psychology*, *223*(1), 3–13. https://doi.org/10.1027/2151-2604/a000194

Blömeke, S., Jenßen, L., Grassmann, M., Dunekacke, S., & Wedekind, H. (2017). Process mediates structure: The relation between preschool teacher education and preschool teachers' knowledge. *Journal of Educational Psychology*, *109*(3), 338–354. https://doi.org/10.1037/edu0000147

Bruns, J., Eichen, L., & Gasteiger, H. (2017). Mathematics-related competence of early childhood teachers visiting a continuous professional development course: An intervention study. *Mathematics Teacher Education and Development*, *19*(3), 76–93.

Burroughs, N., Gardner, J., Lee, Y., Guo, S., Touitou, I., Jansen, K., & Schmidt, W. (2019). Teaching for excellence and equity: Analyzing teacher characteristics, behaviors and student outcomes with TIMSS. *A review of the literature on teacher effectiveness and student outcomes* (pp. 7–17). Cham: Springer.

Claessens, A., Engel, M., & Curran, F. C. (2014). Academic content, student learning, and the persistence of preschool effects. *American Educational Research Journal*, *51*(2), 403–434. https://doi.org/10.3102/0002831213513634

Cohen, J. (1988). *Statistical power analysis for the behavioral sciences* (2nd ed.). Hillside, NJ: Lawrence Erlbaum Associates.

Duncan, G. J., Dowsett, C. J., Claessens, A., Magnuson, K., Huston, A. C., Klebanov, P., Pagani, L. S., Feinstein, L., Engel, M., Brooks-Gunn, J., Sexton, H., Duckworth, K., & Japel, C. (2007). School readiness and later achievement. *Developmental Psychology*, *43*(6), 1428–1446. https://doi.org/10.1037/0012-1649.43.6.1428

Dunekacke, S., Jenßen, L., Eilerts, K., & Blömeke, S. (2016). Epistemological beliefs of prospective preschool teachers and their relation to knowledge, perception, and planning abilities in the field of mathematics: A process model. *ZDM – Mathematics Education*, *48*(1–2), 125–137. https://doi.org/10.1007/s11858-015-0711-6

Engel, M., Claessens, A., Watts, T., & Farkas, G. (2016). Mathematics content coverage and student learning in kindergarten. *Educational Researcher*, *45*(5), 293–300. https://doi.org/10.3102/0013189X16656841

Field, A. (2018). *Discovering statistics using IBM SPSS statistics*. London, UK: Sage.

Gasteiger, H., Bruns, J., Benz, C., Brunner, E., & Sprenger, P. (2020). Mathematical pedagogical content knowledge of early childhood teachers: A standardized situation-related measurement approach. *ZDM – Mathematics Education*, *Online first*. https://doi.org/10.1007/s11858-019-01103-2

Hattie, J. (2009). *Visible learning: A synthesis of over 800 meta-analyses relating to achievement*. London: Routledge.

Kaiser, G., Blömeke, S., König, J., Busse, A., Döhrmann, M., & Hoth, J. (2017). Professional competencies of (prospective) mathematics teachers – Cognitive versus situated approaches. *Educational Studies in Mathematics, 94*(2), 1–22. https://doi.org/10.1007/s10649-016-9713-8

Kersting, N. B., Givvin, K. B., Thompson, B. J., Santagata, R., & Stigler, J. W. (2012). Measuring usable knowledge: Teachers' analyses of mathematics classroom videos predict teaching quality and student learning. *American Educational Research Journal, 49*(3), 568–589. https://doi.org/10.3102/0002831212437853

Kleickmann, T., Richter, D., Kunter, M., Elsner, J., Besser, M., Krauss, S., & Baumert, J. (2013). Teachers' content knowledge and pedagogical content knowledge: The role of structural differences in teacher education. *Journal of Teacher Education, 64*, 90–106. https://doi.org/10.1177/0022487112460398

Klibanoff, R. S., Levine, S. C., Huttenlocher, J., Vasilyeva, M., & Hedges, L. V. (2006). Preschool children's mathematical knowledge: The effect of teacher "math talk". *Developmental Psychology, 42*(1), 59–69. https://doi.org/10.1037/0012-1649.42.1.59

König, J., Blömeke, S., & Kaiser, G. (2015). Early career mathematics teachers' general pedagogical knowledge and skills: Do teacher education, teaching experience, and working conditions make a difference? *International Journal of Science and Mathematics Education, 13*(2), 331–350. https://doi.org/10.1007/s10763-015-9618-5

Krauss, S., Bruckmaier, G., Lindl, A., Hilbert, S., Binder, K., Steib, N., & Blum, W. (2020). Competence as a continuum in the COACTIV-Study: The cascade model. *ZDM – Mathematics Education,52*, 311–332. https://doi.org/10.1007/s11858-020-01151-z

Kunter, M., Klusmann, U., Baumert, J., Richter, D., Voss, T., & Hachfeld, A. (2013). Professional competence of teachers: Effects on instructional quality and student development. *Journal of Educational Psychology, 105*(3), 805–820. https://doi.org/10.1037/a0032583

Lee, J. (2010). Exploring kindergarten teachers' pedagogical content knowledge of mathematics. *International Journal of Early Childhood, 42*(1), 27–41. https://doi.org/10.1007/s13158-010-0003-9

Lee, J., Meadows, M., & Lee, J. O. (2003). *What causes teachers to implement high quality mathematics education more frequently: Focusing on teachers' pedagogical content knowledge* (ED 472 327). Washington, DC: ERIC Clearinghouse on Teaching and Teacher Education.

Lee, J. E. (2017). Preschool teachers' pedagogical content knowledge in mathematics. *International Journal of Early Childhood, 49*(2), 229–243. https://doi.org/10.1007/s13158-017-0189-1

Lehrl, S., Kluczniok, K., & Rossbach, H. G. (2016). Longer-term associations of preschool education: The predictive role of preschool quality for the development of mathematical skills through elementary school. *Early Childhood Research Quarterly, 36*, 475–488. https://doi.org/10.1016/j.ecresq.2016.01.013

Li, X. (2020). Investigating U.S. preschool teachers' math teaching knowledge in counting and numbers. *Early Education and Development, Online first.* https://doi.org/10.1080/10409289.2020.1785226

Li, X., McFadden, K., & DeBey, M. (2019). Is It DAP? American preschool teachers' views on the developmental appropriateness of a preschool math lesson from China. *Early Education and Development, 30*, 765–787. https://doi.org/10.1080/10409289.2019.1599094

McCray, J. S., & Chen, J. Q. (2012). Pedagogical content knowledge for preschool mathematics: Construct validity of a new teacher interview. *Journal of Research in Childhood Education, 26*(3), 291–307. https://doi.org/10.1080/02568543.2012.685123

Melhuish, E., Quinn, L., Sylva, K., Sammons, P., Siraj-Blatchford, I., Taggart, B. (2013). Preschool affects longer term literacy and numeracy: Results from a general population longitudinal study in Northern Ireland. *School Effectiveness and School Improvement*, *24*(2), 234–250. https://doi.org/10.1080/09243453.2012.749796

Oppermann, E., Anders, Y., & Hachfeld, A. (2016). The influence of preschool teachers' content knowledge and mathematical ability beliefs on their sensitivity to mathematics in children's play. *Teaching and Teacher Education*, *58*, 174–184. https://doi.org/10.1016/j.tate.2016.05.004

Piasta, S. B., Logan, J. A. R., Pelatti, C. Y., Capps, J. L., & Petrill, S. A. (2015). Professional development for early childhood educators: Efforts to improve math and science learning opportunities in early childhood classrooms. *Journal of Educational Psychology*, *107*(2), 407–422. https://doi.org/10.1037/a0037621

Piasta, S. B., Pelatti, C. Y., & Miller, H. L. (2014). Mathematics and science learning opportunities in preschool classrooms. *Early Education and Development*, *25*(4), 445–468. https://doi.org/10.1080/10409289.2013.817753

Polly, D., Martin, C. S., McGee, G. R., Wang, C., Lambert, R. G., & Pugalee, D. K. (2017). Designing curriculum-based mathematics professional development for kindergarten teachers. *Early Childhood Education Journal*, *45*, 659–669. https://doi.org/10.1007/s10643-016-0810-1

Qian, H., & Youngs, P. (2016). The effect of teacher education programs on future elementary mathematics teachers' knowledge: A five- country analysis using TEDS-M data. *Journal of Mathematics Teacher Education*, 371–396. https://doi.org/10.1007/s10857-014-9297-0

Schmidt, W. H., Cogan, L., & Houang, R. (2011). The role of opportunity to learn in teacher preparation: An international context. *Journal of Teacher Education*, *62*(2), 138–153. https://doi.org/10.1177/0022487110391987

Schön, D. A. (1983). *The reflective practitioner – How professionals think in action*. New York: Basis Books.

Thornton, J. S., Crim, C. L., & Hawkins, J. (2009). The impact of an ongoing professional development program on prekindergarten teachers' mathematics practices. *Journal of Early Childhood Teacher Education*, *30*(2), 150–161. https://doi.org/10.1080/10901020902885745

Tirosh, D., Tsamir, P., Levenson, E., & Tabach, M. (2011). From preschool teachers' professional development to children's knowledge: Comparing sets. *Journal of Mathematics Teacher Education*, *14*, 113–131. https://doi.org/10.1007/s10857-011-9172-1

Torbeyns, J., Verbruggen, S., & Depaepe, F. (2020). Pedagogical content knowledge in preservice preschool teachers and its association with opportunities to learn during teacher training. *ZDM – Mathematics Education*, *52*(2), 269–280. https://doi.org/10.1007/s11858-019-01088-y

Yang, X., Kaiser, G., König, J., & Blömeke, S. (2020). Relationship between pre-service mathematics teachers' knowledge, beliefs and instructional practices in China. *ZDM – Mathematics Education, Online first.* https://doi.org/10.1007/s11858-020-01145-x

Chapter 3

Age-appropriate performance expectations and learning objectives of early childhood teachers in the field of mathematics

A cross-country comparison of Austria and Switzerland

Karoline Rettenbacher, Lars Eichen, Manfred Pfiffner,
Catherine Walter-Laager

Introduction

Children entering early childhood education and care (ECEC) institutions already have different levels of knowledge and skill in specific educational areas (Clements & Sarama, 2014; Klibanoff, Levine, Huttenlocher, Vasilyeva, & Hedges, 2006). The mathematical knowledge children have before they enter school influences their later mathematical school achievements as well as their later understanding of mathematical concepts as adults (Watts, Duncan, Siegler, & Davis-Kean, 2014). Receiving a high-quality mathematical education at an early age has a positive influence on a child's entire educational path (Sylva, Melhuish, Sammons, Siraj-Blatchford, & Taggart, 2011).

The implementation of high-quality (mathematical) play requires pedagogical activities that are matched to children's needs and interests and prior knowledge (Clements & Sarama, 2014; Fuson, Clements, & Sarama, 2015; Walter-Laager & Fasseing, 2017).

Educational activities need to be planned by observing and assessing children's development and identifying the next developmental steps that lie in the *zone of proximal development* (Vygotskij, 2002) educational offers can be planned accordingly (Walter-Laager & Fasseing, 2017). These requirements place high demands on early childhood teachers and this situation is made even more demanding by the relative dearth of suitable data on children's mathematical development.

Studies on when children acquire certain mathematical abilities and knowledge are quite scarce. Most of the relevant research pertains to the content area *numbers and operations*. Empirical data are also available on content areas such as *patterns and structures, shapes and space, measurement*, and *data, frequencies, and probabilities*, but to a much smaller extent (Benz, Peter-Koop, & Grüßing, 2015; Clements & Sarama, 2014). The research literature tends to contain

DOI: 10.4324/9781003172529-3

generalizing or specific statements based on different research questions. This, together with the relatively high number of omissions or discontinuities, makes it difficult to gain an overall picture.

Early childhood educational programmes and daily pedagogical practice are closely linked to cultural values and convictions (Hammer & He, 2016). MacDonald and Murphy (2019) show that early childhood teachers tend to underestimate the mathematical abilities of young children and that cultural and contextual differences often influence their views on children's capabilities.

Curricular performance expectations and learning objectives

In Austria and the German-speaking part of Switzerland,[1] ECEC institutions include preschools that in both countries are referred to as kindergarten and nurseries. In Switzerland, children aged 3 months to 4 years can attend nurseries. There is a slight difference in Austria, where nursery children are aged 0–3 years, with the entry age varying according to local regional regulations. In Switzerland, kindergarten attendance is compulsory and free of charge from the age of 4 years in most cantons. In Austria, children between the ages of 3 and 6 years attend kindergarten, with the last year being both compulsory and free of charge.

The teacher training in the two countries differs. Early childhood teachers in Austria attend a vocational upper-secondary school or a college. This, then, qualifies them to work in Austrian ECEC institutions (https://www.bmbwf. gv.at/en/Topics/Early-childhood-education/eche_career.html). In contrast, in Switzerland early childhood teacher training is on tertiary level (Eggenberger, 2008). Studies show that both the stage (in-service, pre-service, years of work experience) and the kind of professional development can influence the teacher's resulting professional competences (Blömeke, Jenßen, Grassmann, Dunekacke, & Wedekind, 2017; Hepberger, Lindermeier, Optiz, & Heinze, 2017; Lee, 2017).

A comparison of the curriculum in Austria and Switzerland shows a rather uniform picture of educational objectives in early mathematics, even though the two education systems do display certain structural differences.

The national curriculum *Lehrplan 21* in Switzerland is currently being introduced for the kindergarten- and the primary school level. It is divided into three cycles. Points of reference within the first cycle are used to describe which competence levels in each educational area the kindergarten needs to cover. The competences for cycle 1, for example, are to "understand and use the terms plus, minus, is equal to and the symbols +, −, =" (p. 216) or to "can show how they count" (p. 224), and they apply to children up to the third grade of primary school (Bildungsdirektion des Kantons Zürich, 2017). In comparison to the Austrian national curriculum, the *Lehrplan 21* provides detailed specifications for learning objectives.

Austria's national curriculum applies to all ECEC institutions with children aged 0–6 years in their care. Which content area or specific abilities children must acquire in the area of early mathematical education is not explicitly determined (Charlotte-Bühler-Institut, 2009).

Age-appropriate performance expectations in early mathematics education

Early childhood teacher's competences enable them to create learning opportunities based on professional decisions. Their knowledge concerning children's *zone of proximal development* enables them to devise or plan different types of learning opportunities. There are currently three competence models, in the German-speaking countries, under discussion concerning the demands facing early childhood teachers in mathematics (Blömeke, Gustafsson, & Shavelson, 2015; Gasteiger & Benz, 2016; Lindmeier, 2011). They may all be used to illustrate how different facets of competence interact. In all three models, the early childhood teacher's knowledge about the age-appropriate development of children's mathematical abilities and understanding plays an important role in MPCK (mathematical pedagogical content knowledge [MPCK]). Unfortunately, however, there is also a current lack of empirical data concerning age-appropriate performance expectations (Browne & Richard Wong, 2017).

In the competence model of Blömeke et al. (2015), competence itself is understood as a continuum, MPCK is seen as one of three dimensions of knowledge, and is itself made up of four sub-dimensions. Two of these sub-dimensions are knowledge about the development of mathematical abilities in children aged 3–6 years[2] and knowledge about how these abilities can be diagnosed (Dunekacke, Jenßen, Eilerts, & Blömeke, 2016; Jenßen, Dunekacke, Gustafsson, & Blömeke, 2019).

In the structural model of Lindmeier (2011), basic knowledge (BK) comprises the content knowledge (CK) and PCK. This thus concerns both knowledge of early mathematical content, structures, and working methods, as well as knowledge of how this can be qualitatively taken up and applied in teaching practice in order to facilitate the mathematical development of children and their transition into primary school. It is assumed that performance expectations are part of BK, as it includes the CK and PCK about the facilitation of children's mathematical competences and knowledge about children's development (Hepberger et al., 2017).

The third model is a structural process model. This attempts to make various structural facets of knowledge more visible, and to evaluate how they may interact, for example the interaction between the situational observation and perception (SOP) with the pedagogical didactic action. Gasteiger and Benz (2016) distinguish between two types of knowledge; explicit and implicit knowledge. The former includes knowledge of mathematical content, concepts, and appropriate choice of methods and materials, and the ability to recognize the developmental processes in mathematical abilities. Implicit knowledge is experience-based (Gasteiger & Benz, 2016).

Action planning as one facet of competence

Action planning is another facet of the early childhood teacher's MPCK that is included in all three above-mentioned models of professional competence (Blömeke et al., 2015; Gasteiger & Benz, 2016; Lindmeier, 2011). Early childhood teachers plan activities based on their professional knowledge and on the observations they make. Learning objectives when planning learning activities are always oriented towards general objectives (e.g., helping the child attain self-competence, social competence, or technical competence), the topic, and type of activity (Walter-Laager & Fasseing, 2017).

In the model devised by Blömeke et al. (2015), the competence facet MCK (mathematical content knowledge) forms the basis for further consideration of how a child's mathematical knowledge may be facilitated. During action planning, decisions on learning objectives and their didactic implementation are based on an interpretation of the corresponding observations. The teacher's PCK can predict how situations are perceived and thus actions are adequately planned (Dunekacke et al., 2016; Dunekacke, Jenßen, & Blömeke, 2015). Dunekacke et al. (2015) examined the connection between MCK and the teachers' situational perceptions and action planning skills with a paper-pencil test and a vignette test for $N = 354$ early childhood teachers. They report a moderately predictive effect of MCK on situational perception, with the latter being a strong predictor of action planning. However, no direct effect could be found for MCK on action planning. Thus, MCK seems to be a necessary but not sufficient prerequisite for action planning (Dunekacke et al., 2015).

In Lindmeier's structural model (2011), action planning is located within reflexive competence. This includes the ability and willingness to prepare and evaluate mathematical activities with respect to the competence development on the basis of BK. The focus lies on preparing specific learning opportunities adapted to a difficulty level appropriate for the children (Lindmeier, 2011).

In the structure process model of Gasteiger and Benz (2016), the SOP includes the ability of action planning as well as to spontaneously offer appropriate stimuli in order to support the child's learning. Being able to perceive both mathematical content and children's abilities in natural learning situations is a prerequisite for early childhood teachers attempting to support that mathematical learning. The ability to recognize the child's learning stage and to react suitably is part of PCK and manifests itself, for example, in the formation of diagnostic questions or in the targeted selection of learning stimuli (Gasteiger & Benz, 2016).

As a rule, the actual learning progress of children has to be evaluated frequently. In the school context, this is done either through tests or through a process of observation (Lindmeier, Heinze, & Reiss, 2013). The learning objectives, set by early childhood teachers, are oriented towards the desired learning outcome or progress. The learning objectives need to be manageable and yet challenging for the child. To ensure that learning objectives are appropriate, and to provide suitable orientation, benchmarks of expected performance in certain age groups or rather, developmental groups are derived on the basis of empirical data (Shapiro, 2011). These benchmarks are first deduced based on the age or developmental stage of the children and then used to formulate performance expectations.

Bruns (2014) reports that adaptive fostering behaviour in early mathematics is only marginally pronounced among early childhood teachers in Germany. She suspects that the relative scarcity of activities within the zone of proximal development is connected to a lack of general pedagogical knowledge, and a lack of CK, on the part of the early childhood teacher. Engel, Claessens and Finch (2013) report that early childhood teachers often address mathematical content in kindergarten that is too easy for the majority of the children in their classes. The assumption is that early childhood teachers feel uneasy when teaching mathematical content and therefore resort to simpler mathematical concepts. This has the unintended result that only a small proportion of children are adequately supported in their learning (Engel et al., 2013). Of course, it is also possible that early childhood teachers consider that such "unchallenging" learning objectives are actually adequate for the respective age groups.

Research question

Both theoretical considerations and empirical data show how important the professional (mathematical) competences of early childhood teachers are for action planning and the implementation of qualitative learning opportunities. The following research questions are thus addressed:

1   How accurate are the age-appropriate performance expectations of early childhood teachers in Switzerland and Austria for children aged 3–6 years?
2   What learning objectives do early childhood teachers in Switzerland and Austria set when preparing a typical math activity in the content areas *measurement,* and *shapes and space*?
3   How do early childhood teachers in Switzerland and Austria make sure that the children reach their set learning objectives?

Method

The data from Austria and Switzerland were collected as part of the *BELMI 3-6* project. The project compares the educational goals (Keller, 2011), learning objectives, and age-appropriate performance expectations of early childhood teachers and the structure of math activities in Austria, China, Vietnam, Switzerland, and the USA.

Table 3.1 Sample item from the questionnaire on age-appropriate performance expectations

At which age do you **expect** children to show the following skills? Please mark with one single cross.							
Years; months	*1.0–1.6*	*1.7–2.0*	*2.1–2.6*	*2.7–3.0*	*3.1–4.0*	*4.1–5.0*	*5.1–6.0*
(Example: 1; 0 = 1 year/1;							
6 = 1 year, 6 months)							
Expected skills of the							
children							
The child cuts round and angular shapes	☐	☐	☐	☐	☒	☐	☐

Data collection

For the age-appropriate performance expectations (research question 1), the early childhood teachers were asked to rate items on when children achieve specific mathematical abilities in different mathematical content areas (Benz et al., 2015; Clements & Sarama, 2014) on the basis of age cohorts between 1 and 6 years. In this chapter, results from the content areas *shapes and space* and *measurement* will be presented. A sample item can be found in Table 3.1.

The statements on mathematical abilities are taken from the observation tool KiDiT® (Walter-Laager, Pfiffner, & Schwarz, 2012). Information on age-cohort accuracy was derived from various empirical studies and compared to empirical data from the KiDiT database in an attempt to close the existing gaps in knowledge.

An explorative approach was chosen in order to capture the design of math activities (research questions 2 and 3). A sample of 12 early childhood teachers were asked to prepare a math activity they would usually offer on an ordinary day, and this was then videotaped. The math activities were prepared for 3–4-year-old children in the content area *shapes and space*, and for 5–6-year-old children in the content area *measurement*. The choice of these mathematical content areas was quite deliberate. Empirical research suggests that while early childhood teachers feel comfortable when offering activities in the content area *numbers and operations*, this is apparently not the case for the areas *measurement* and *shapes and space* which still remain relatively underrepresented. After the observed and recorded math activity, a structured reflective interview was then conducted with the early childhood teachers to assess the chosen learning objectives and form of evaluation of the children's learning progress.

Sample

An ad-hoc sample of $N = 699$ early childhood teachers from Austria and Switzerland filled out the questionnaire about age-appropriate performance expectations in early mathematics (see also table 3.2).

Table 3.2 Sample of early childhood teachers in
Austria and Switzerland

	Austria (n)	Switzerland (n)
Pre-service	91	427
In-service	86	95
Total	177	522

Of the Austrian early childhood teachers, 12.8% have an academic degree, and the other 87.2% have completed vocational school training. In the Swiss sample, 16.8% had completed vocational education and training, 52.7% have a bachelor's or master's degree, and 25.3% have some other form of degree. The average age of the Austrian early childhood teachers is 40.3 years, in the Swiss sample it is 38.9 years.

To collect data on the structuring of math activities, a regional ECEC provider in Austria recruited a list of potential participants among early childhood teachers during a meeting with kindergarten directors. These early childhood teachers were then contacted per email or phone and asked to participate for the interview and videotaping. In Switzerland, a list of potential contacts was drawn up by the authors of the present chapter. These were then contacted by email. A high degree of willingness among participants can thus be assumed for both countries. In three groups of 5–6-year-old children and four groups of 3–4-year-old children, videos of everyday math activities were recorded in Austria. In Switzerland, three groups of 5–6-year-old children and three groups of 3–4-year-old children took part. The structured reflective interviews with altogether 12 early childhood teachers were then conducted right after the filming. The seven Austrian early childhood teachers interviewed had all completed their vocational training as kindergarten teachers and had between 1 and 15 years of work experience. The five Swiss early childhood teachers interviewed had all completed a bachelor's degree during their training as kindergarten teachers and had been working in the field for 5–27 years.

Analysis

Various statistical procedures were used to analyse the data on age-appropriate performance expectations. This included the use of descriptive statistics and non-parametric tests (Mann–Whitney U). The aim of this analysis was to determine differences between pre-service and in-service early childhood teachers, differentiated for Austria versus Switzerland. Results of this quantitative analysis represent the basis for the following qualitative analyses.

The content employed in the reflective interview was theory based and determined prior to the interview. This entailed the use of structured qualitative content analysis (Mayring, 2015). Such an analysis allows for a theory-guided, systematic analysis of conversations and allows for the formation of deductive categories on the basis of content structure. The resulting material was analysed using the software MAXQDA. For the compilation of the results, the text passages from the transcripts were sorted into categories, paraphrased, and each category (subcategories and main categories) was summarized individually (Mayring, 2015).

Results

The following section presents the results obtained for each of the research questions.

Age-appropriateness of performance expectations (research question 1)

The bars in Figures 3.1 and 3.2 show the percentage of the early childhood teachers from Austria and Switzerland that rated the age-appropriate performance expectations accurately, that is in accordance with the age specifications given in the research literature. A close look at the results reveals that for a large proportion of early childhood teachers, in both countries, expectations were often highly inaccurate.

To find out whether the group differences are significant and owing to the partial lack of a normal distribution among the variables, non-parametric tests were performed. The results of the Mann–Whitney U test and their significance levels are presented in Table 3.3.

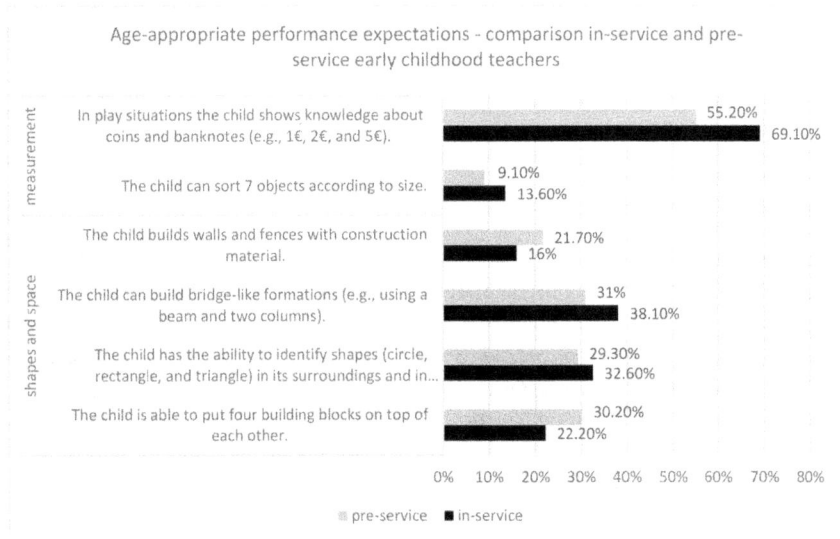

Figure 3.1 Percentage of accurate age-appropriate performance expectations of preservice ($n = 518$) and in-service ($n = 182$) early childhood teachers.

Age-appropriate performance expectations - country comparison

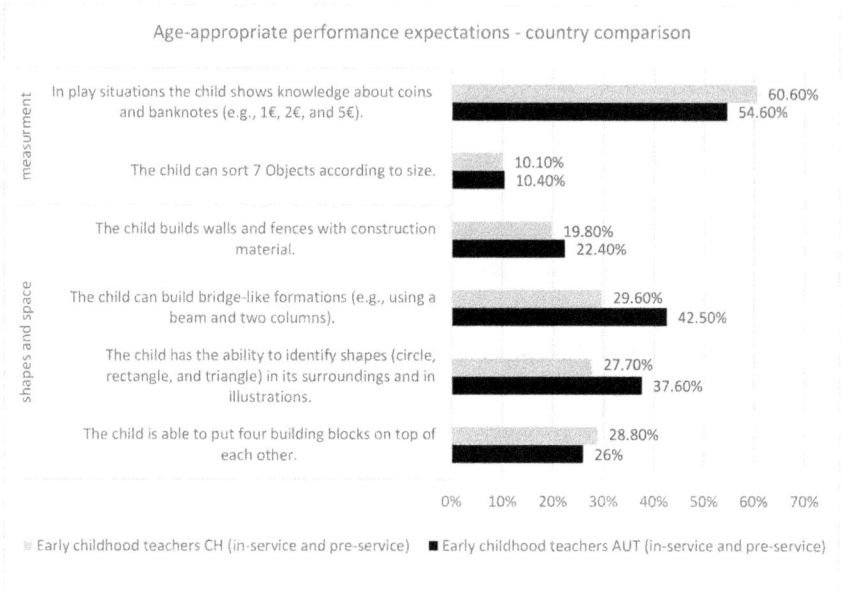

Figure 3.2 Percentage of accurate age-appropriate performance expectations of early childhood teachers in a country comparison (AUT *n* = 177; CH *n* = 522).

For the results in Austria, a significant mean difference between the two groups (pre-service; in-service) is found for only two items, together with low effect sizes (Cohen, 1992). In Switzerland five of the six items show a significant group difference, together with low effect sizes (Cohen, 1992), indicating that educational level in Switzerland has a more significant effect on how early childhood teachers rate performance expectations.

Learning objectives (research question 2)

The following section provides an overview and a description of the interview data, analysed with the content analysis, about the learning objectives in the cross-country comparison. The mathematical learning objectives are divided into the content areas *measurement, shapes and space,* and *other objectives.* The category *process-related competences* was also included but did not provide any results. The process-related competences, such as problem solving or arguing, span across all mathematical content areas (Benz et al., 2015; Clements & Sarama, 2014).

Measurement

Two of the Austrian and all three Swiss early childhood teachers indicated in the interview that they chose *length* as the topic for the math activity (see table 3.4).

Table 3.3 Percentage data for accurate age-appropriate performance expectations by country (AUT; CH) and educational status (pre-service; in-service). Significant results are highlighted in bold.

Items	AUT			CH		
	Pre-service (%)	In-service (%)	Mann–Whitney U test	Pre-service (%)	In-service (%)	Mann–Whitney U test
In play situations, the child shows knowledge about coins and banknotes (e.g., 1€, 2€, and 5€)	51.10	58.50	$U = 3869.00, p = .539, r = .61$	56.10	78.50	**$U = 24,109.00, p < .001, r = .19$**
The child can sort 7 objects according to size	5.60	15.90	**$U = 4743.50, p < .001, r = .28$**	9.80	11.70	**$U = 23,071.50, p = .005, r = .12$**
The child builds walls and fences with construction material	26.70	18.30	$U = 3941.00, p = .415, r = .82$	20.60	14.00	$U = 20,896.00, p = .224, r = 1.22$
The child can build bridge-like formations (e.g., using a beam and two columns)	45.60	39.00	$U = 3999.00, p = .316, r = 1.00$	27.90	37.20	**$U = 22,343.50, p = .035, r = .09$**
The child has the ability to identify shapes (circle, rectangle, and triangle) in its surrounding environment and in illustrations	37.10	36.60	**$U = 4666.50, p = .002, r = .24$**	27.60	29.00	**$U = 24,681.00, p < .001, r = .18$**
The child is able to put four building blocks on top of each other	25.80	26.80	$U = 3905.00, p = .370, r = .89$	31.10	18.10	**$U = 23,438.00, p = .003, r = .13$**

Note: Significant results are highlighted in bold.

Table 3.4 Learning objectives of early childhood teachers from Austria and Switzerland in the content area measurement

Austria	Switzerland
• Understanding size relations • Getting to know the meter stick and the scale as standardized measuring instruments • The measurements as the number of units of measurement • Measuring with non-standard measuring instruments (e.g., with a piece of string) • Size comparisons based on one's own height and the height of the other children in the group • Understanding the measuring process • Understand the relational terms heavy and light, bigger than, and smaller than	• Knowledge about the meaning of the terms: equally long, shorter, and longer • Understanding and executing a measurement using non-standard measuring instruments (e.g., a piece of string, cubes) • Understanding units of measurement of non-standard measuring instruments • Understanding that measuring is the counting of units of measure • Comparing sizes

One Austrian early childhood teacher decided to choose *weight* as a topic. The statements of the early childhood teachers concerning content are similar for both countries. The early childhood teachers stated the following learning objectives.

Shapes and space

In total, six early childhood teachers prepared a math activity for the content area shapes and space for children aged 3–4 years. Five early childhood teachers envisioned that the children would be able to recognize and name the introduced geometric objects or shapes (circle) in everyday objects (e.g., in form of a building block or a plate). One early childhood teacher explicitly concentrated on only one shape, the circle, as a learning objective in the math activity.

The early childhood teachers stated the following learning objectives found in table 3.5.

Table 3.5 Learning objectives of early childhood teachers from Austria and Switzerland in the content area shapes and space

Austria	Switzerland
• Recognizing and naming the basic geometric shapes: triangle, square, and circle • Being able to recognize and name the geometric shapes or the shape (circle) in everyday objects (e.g., in a building block or plate)	• Recognition and naming of geometric shapes (triangle, square, circle, hexagon) • Recognizing and naming the properties of the shapes (triangle, square, circle, hexagon) • Ability to recognize and name the geometric shapes in surfaces of everyday objects (e.g., in a building block or plate)

Other goals

Compared to the Swiss early childhood teachers, the Austrian early childhood teachers also stated other objectives in addition to their mathematical learning objectives. When questioned about their mathematical learning objectives, four of the Austrian early childhood teachers also stated objectives that were not math-specific but that were assigned to the category *other objectives*, such as the need to strengthen a child's social-emotional competence, child's enjoyment of an activity, or facilitating fun in natural sciences. The Swiss early childhood teachers did not name any learning objectives in this category.

Evaluation of the learning objectives (research question 3)

In addition to the question about learning objectives, the early childhood teachers in both countries were asked how they established whether children achieve their set learning objectives.

In Austria, the early childhood teachers interviewed stated that they determined learning progress mostly through observation of the children. Four of the early childhood teachers reported that they recognized learning progress when children carried out the activities as intended by the early childhood teachers, for example, by placing little glass stones in a circle on a pre-drawn line, colouring shapes on a worksheet or naming shapes they recognize in everyday objects. Three of the early childhood teachers did not provide any detailed information on how they recognize a learning progress in children. They merely stated, for example, that the length of time children spent on an activity and the child's enthusiasm for the activity were indicators of learning progress. Another early childhood teacher described the moment in which the children learn by saying that "a light goes on in their head" and they suddenly understand. The teacher, however, did not go into any detail about how this may become observable. One of the early childhood teachers interviewed stated that she could not observe any learning progress immediately after the activity. She argued that children need some time to process what they had learned and that a learning progress could only be ascertained at some later point, for example, everyday conversations.

More specific information is provided by the early childhood teachers from Switzerland. Four of the five Swiss early childhood teachers stated here that they can determine learning progress by observing the children's behaviour, by a child's use of correct mathematical terms, or when children are able to explain why they chose a certain selection of geometric shapes or the correct use of measuring instruments.

Discussion

The results of this study complement the results described by other authors that teacher training has an effect on competence (Hepberger et al., 2017; Lee, 2017). Our results show significant group differences in performance expectations, but not a country difference per se. The results are inconsistent: A rather large percentage of early childhood teachers exhibited inaccurate performance expectations, that is they were either too high or too low-performance expectations. The cross-country difference that emerged between in-service early childhood teachers supports the view that the type of teacher training experienced may play a role in expectation formation. However, it is also clear that there are other factors that influence the performance expectations of early childhood teachers, such as the existence of different curricular requirements. For example, in the case of Switzerland and Austria, large structural differences prevail in the formulation of children's competences.

The importance of accurate formation of performance expectations becomes clear once one focuses on the role of action planning in early childhood teaching. Early childhood teachers observe the whole group of children in order to determine which developmental steps are pending and whether these may be applied homogeneously for the whole group, or whether implementation needs to be differentiated (Walter-Laager & Fasseing, 2017). Such observations and considerations provide a starting point for the planning of educational activities (Clements & Sarama, 2014; Meyer & Walter-Laager, 2012). For the teacher, expectations provide a reference value when deciding upon the further developmental steps which need to be taken. Thus, in order for early childhood teachers to assess *zones of proximal development* correctly, it is essential that their performance expectations be as accurate as possible.

The learning objectives set in pedagogical activities are based on observations as well as on internalized performance expectations. These are then matched to the needs of the individual child, the small group, subgroup or the whole group and implemented appropriately as the case may be (Walter-Laager & Fasseing, 2017). The results described in the present study show that the learning objectives of the early childhood teachers in both countries are, on the whole, very similar. One difference was identified in that the Austrian early childhood teachers additionally mentioned general pedagogical objectives among their learning activities, for example the need for children to have fun. The Swiss early childhood teachers did not mention any such objectives and their objectives were only stated in terms of mathematical content. In addition, in terms of evaluating learning objectives, the views expressed by the Swiss early childhood teachers appear to be more distinct than those of their Austrian counterparts.

Conclusion

Given the present dearth of reliable empirical data concerning the mathematical development of young children, one should not expect too much of early childhood teachers. However, in a more positive vein, this lack of data may also be seen as a clarion call for future research. There is now clearly a need for more empirical evidence on the development of children's competencies in almost all mathematical content areas. With a view to improving professional development, especially in Austria, and in order to understand how learning objectives are determined and implemented, it would appear necessary to examine in much greater detail the formation and interplay of mathematical CK and PCK among early childhood teachers.

Limitations of the study

The sample used for the questionnaire was not a random sample but an ad-hoc sample. There was also obvious (self-)selection bias concerning the interview partners (given their high level of willingness to participate in the study). It should also be noted that the benchmarks used to determine the accuracy of the age cohorts for the sample items on performance expectations need to be looked at critically. Additionally, as mentioned in the theoretical part of this chapter, empirical data on the mathematical development of children are relatively scarce. Even though the KiDiT data stem from a large database, distortions cannot be ruled out. Nor can they be ruled out with respect to the ratings provided by the early childhood teachers. Nonetheless, the data described here serve to provide further insight into the relationship between of age-appropriate performance expectations and learning objectives in early childhood education.

Notes

1 Which is the region of Switzerland we will refer to in this chapter.
2 It should be noted here that children develop mathematical skills even before the age of three (e.g., Starkey, Spelke, & Gelman, 1990).

References

Benz, C., Peter-Koop, A., & Grüßing, M. (2015). Frühe mathematische Bildung: Mathematiklernen der Drei- bis Achtjährigen [Early mathematical education: Learning mathematics at the age of three to eight]. *Mathematik Primarstufe und Sekundarstufe I+ II*. Berlin, Heidelberg: Springer Spektrum. https://doi.org/10.1007/978-3-8274-2633-8

Bildungsdirektion des Kantons Zürich (2017). Lehrplan für die Volksschule des Kantons Zürich auf der Grundlage des Lehrplans 21 [Curriculum for primary school of the canton Zurich on the basis of the curriculum 21]. Retrieved September 24, 2018, from https://zh.lehrplan.ch/container/ZH_Grundlagen.pdf

Blömeke, S., Gustafsson, J.-E., & Shavelson, R. J. (2015). Beyond dichotomies: Competence viewed as a continuum. *Zeitschrift Fuer Psychologie, 223*(1), 3–13. https://doi.org/10.1027/2151-2604/a000194

Blömeke, S., Jenßen, L., Grassmann, M., Dunekacke, S., & Wedekind, H. (2017). Process mediates structure: The relation between preschool teacher education and preschool teachers' knowledge. *Journal of Educational Psychology, 109*(3), 338–354. https://doi.org/10.1037/edu0000147

Browne, L., & Richard Wong, K.-S. (2017). Transnational comparisons of teacher expectation of mathematical functional ability in early years and key stage 1 pupils – A study undertaken in Hong Kong and England. *Education, 45*(4), 504–515. https://doi.org/10.1080/03004279.2016.1140798

Bruns, J. (2014). *Adaptive Förderung in der elementarpädagogischen Praxis: Eine empirische Studie zum didaktischen Handeln von Erzieherinnen und Erziehern im Bereich Mathematik* [Adaptive promotion in elementary pedagogical practice: Empirical study on the didactic actions of educators in the field of mathematics]. Empirische Studien zur Didaktik der Mathematik. Münster: Waxmann.

Charlotte-Bühler-Institut (2009). *Bundesländerübergreifender BildungsRahmenPlan für elementare Bildungseinrichtungen in Österreich: Endfassung, August 2009* [National curriculum for early childhood institutions across all provinces in Austria]. Wien: BMUKK. Retrieved September 24, 2018, from http://media.obvsg.at/AC11757249-2001

Clements, D. H., & Sarama, J. (2014). *Learning and teaching early math: The learning trajectories approach* (2nd ed.). Studies in mathematical thinking and learning series. New York, NY: Routledge.

Cohen, J. (1992). Statistical power analysis. *Psychological Bulletin, 112*(1), 155–159. https://doi.org/10.1111/1467-8721.ep10768783

Dunekacke, S., Jenßen, L., & Blömeke, S. (2015). Effects of mathematics content knowledge on pre-school teachers' performance a video-based assessment of perception and planning abilities in informal learning situations. *International Journal of Science and Mathematics Education, 13*(2), 267–286. https://doi.org/10.1007/s10763-014-9596-z

Dunekacke, S., Jenßen, L., Eilerts, K., & Blömeke, S. (2016). Epistemological beliefs of prospective preschool teachers and their relation to knowledge, perception, and planning abilities in the field of mathematics: A process model. *ZDM: The International Journal on Mathematics Education, 48*(1), 125–137. https://doi.org/10.1007/s11858-015-0711-6

Eggenberger, D. (2008). Ausbildung von Fachleuten in der familienergänzenden Kinderbetreuung: Aktuelle Situation in der deutschen Schweiz [Training of professionals in supplementary childcare: Current situation in German-speaking Switzerland]. Retrieved November 12, 2020, from https://paeda-logics.ch/wp-content/uploads/2014/10/ausbildungssituation.pdf

Engel, M., Claessens, A., & Finch, M. A. (2013). Teaching students what they already know? The (mis)alignment between mathematics instructional content and student knowledge in kindergarten. *Educational Evaluation and Policy Analysis, 35*(2), 157–178. https://doi.org/10.3102/0162373712461850

Fuson, K. C., Clements, D. H., & Sarama, J. (2015). Making early math education work for all children. *Phi Delta Kappan, 97*(3), 63–68. https://doi.org/10.1177/0031721715614831

Gasteiger, H., & Benz, C. (2016). Mathematikdidaktische Kompetenz von Fachkräften im Elementarbereich – ein theoriebasiertes Kompetenzmodell [Professional competence of early childhood educators in mathematics education – A theory based competence model]. *Journal für Mathematik-Didaktik, 37*, 263–287. https://doi.org/10.1007/s13138-015-0083-z

Hammer, A. S. E., & He, M. (2016). Preschool teachers' approaches to science: A comparison of a Chinese and a Norwegian kindergarten. *European Early Childhood Education Research Journal, 24*(3), 450–464. https://doi.org/10.1080/1350293X.2014.970850

Hepberger, B., Lindermeier, A., Optiz, E. M., & Heinze, A. (2017). "Zähl' nochmal genauer!" – Handlungsnahe mathematikbezogene Kompetenzen von pädagogischen Fachkräften erheben ["Count more precisely again" – Action-oriented math-related competencies of kindergarten teachers]. In S. Schuler, C. Streit, & G. Wittmann (Eds.), *Perspektiven mathematischer Bildung im Übergang vom Kindergarten zur Grundschule* (pp. 239–253). Wiesbaden: Springer Spektrum. https://doi.org/10.1007/978-3-658-12950-7

Jenßen, L., Dunekacke, S., Gustafsson, J.-E., & Blömeke, S. (2019). Intelligence and knowledge: The relationship between preschool teachers' cognitive dispositions in the field of mathematics. *Zeitschrift Für Erziehungswissenschaft, 22*(6), 1313–1332. https://doi.org/10.1007/s11618-019-00911-2

Keller, H. (2011). *Kinderalltag: Kulturen der Kindheit und ihre Bedeutung für Bindung, Bildung und Erziehung* [Everyday life of children: Childhood cultures and their significance for attachment, education and upbringing]. Berlin: Springer. https://doi.org/10.1007/978-3-642-15303-7

Klibanoff, R. S., Levine, S. C., Huttenlocher, J., Vasilyeva, M., & Hedges, L. V. (2006). Preschool children's mathematical knowledge: The effect of teacher "math talk". *Developmental Psychology, 42*(1), 59–69. https://doi.org/10.1037/0012-1649.42.1.59

Lee, J. E. (2017). Preschool teachers' pedagogical content knowledge in mathematics. *International Journal of Early Childhood, 49*(2), 229–243. https://doi.org/10.1007/s13158-017-0189-1

Lindmeier, A. (2011). Modeling and measuring knowledge and competencies of teachers: A threefold domain-specific structure model for mathematics. *Empirische Studien zur Didaktik der Mathematik* (Vol. 7). Münster: Waxmann.

Lindmeier, A., Heinze, A., & Reiss, K. (2013). Eine Machbarkeitsstudie zur Operationalisierung aktionsbezogener Kompetenz von Mathematiklehrkräften mit videobasierten Maßen [Measuring action-related competences of mathematics teachers with video-based items: A feasibility study]. *Journal Für Mathematik-Didaktik, 34*, 99–119. https://doi.org/10.1007/s13138-012-0046-6

MacDonald, A., & Murphy, S. (2019). Mathematics education for children under four years of age: A systematic review of the literature. *Early Years: An International Journal of Research and Development*. https://doi.org/10.1080/09575146.2019.1624507

Mayring, P. (2015). *Qualitative Inhaltsanalyse: Grundlagen und Techniken* [Qualitative content analysis: Basics and techniques] (12th ed.). Weinheim: Beltz. https://doi.org/10.1007/978-3-531-92052-8_42

Meyer, H., & Walter-Laager, C. (2012). *Leitfaden für Lehrende in der Elementarpädagogik* [Guide for teachers in elementary education]. Berlin: Cornelsen.

Shapiro, E. S. (2011). Best practices in setting progress monitoring goals for academic skill improvement. In A. Thomas & J. Grimes (Eds.), *Best practices in school psychology V* (5th ed., pp. 141–157). Bethesda, MD: NASP.

Starkey, P., Spelke, E. S., & Gelman, R. (1990). Numerical abstraction by human infants. *Cognition, 36*(2), 97–127. https://doi.org/10.1016/0010-0277%2890%2990001-Z

Sylva, K., Melhuish, E., Sammons, P., Siraj-Blatchford, I., & Taggart, B. (2011). Pre-school quality and educational outcomes at age 11 low quality has little benefit. *Journal of Early Childhood Research, 9*(2), 109–124. https://doi.org/10.1177/1476718X10387900

Vygotskij, L. S. (2002). *Denken und Sprechen: Psychologische Untersuchungen* [Thinking and talking]. Weinheim: Beltz.

Walter-Laager, C., & Fasseing, C. (2017). *Kindergarten: Grundlagen aktueller Kindergartendidaktik* [Kindergarten: Basics of current kindergarten didactics]. Winterhur: ProKiga-Lehrmittelverlag.

Walter-Laager, C., Pfiffner, M., & Schwarz, J. (2012). Beobachten mit KiDiT® – Von der Krippe bis zur Schule: Ein webbasiertes Beobachtungswerkzeug für freie Notizen und systematische Beobachtungen in verschiedenen Bildungsbereichen [Observing with KiDiT® – From the crèche to the school. A web-based observation tool for free notes and systematic observations in different educational areas]. *Zeitschrift Frühe Bildung, 1*(3), 165–167. https://doi.org/10.1026/2191-9186/a000050

Watts, T. W., Duncan, G. J., Siegler, R. S., & Davis-Kean, P. E. (2014). What's past is prologue relations between early mathematics knowledge and high school achievement. *Educational Researcher, 43*(7), 352–360. https://doi.org/10.3102/00131 89X14553660

Chapter 4

How pre-service teacher training changes prospective ECEC teachers' emotions about mathematics

Oliver Thiel

Introduction

Mathematics is and has always been important in science, business, engineering, and everyday life (cf. Remmert, 2004). People with mathematical competencies are needed in our society. Therefore, governments are worried when students' results measured by the Programme for International Student Assessment (PISA) (OECD, 2019) stagnate or decline. Many studies show that early childhood mathematics achievement is a strong predictor of success in future school mathematics, other school subjects, and life itself (Carmichael, MacDonald, & McFarland-Piazza, 2014; Duncan, Dowsett, Claessens, Magnuson, & Huston, 2007; Geary, Hoard, Nugent, & Bailey, 2013). As a result, governments encourage early childhood professionals to engage with their children in mathematics learning. For example, the Norwegian government promotes mathematics and science education in Early Childhood Education and Care (ECEC) institutions and primary schools with a national strategy (Norwegian Ministry of Education and Research, 2015). The strategy shall mobilise, raise awareness, and obligate those who work with children to contribute that children learn and explore science and mathematics with motivation and joy (p. 9). One of the objectives is that teachers' competencies in science and mathematics shall be improved (p. 11).

One reason for rather low mathematics competencies might be that many people dislike or even hate this subject (Larkin & Jorgensen, 2016). Emotions and attitudes towards a subject develop from experiences. In a dynamic interaction with the environment, attitudes guide approach to and avoidance of the subject (Metje, Frank, & Croft, 2007). People often persist in negative attitudes and emotions because of a lack of experiences that contradict their prior experiences (Eiser, Fazio, Stafford, & Prescott, 2003). Early experiences of failure can lead to fear that is difficult to dispel later (Bandura, 1977). Therefore, it is an aim in many ECEC curricula that children shall have positive experiences with mathematics. The Norwegian framework plan for the content and tasks of kindergartens states, 'By engaging with quantities, spaces and shapes, kindergartens shall enable the children to ... find pleasure in mathematics', and staff shall

DOI: 10.4324/9781003172529-4

'encourage the children to be curious, find pleasure in mathematics and take an interest in mathematical relationships' (Norwegian Directorate for Education and Training, 2017, pp. 53–54).

To facilitate children's positive experiences with mathematics, teachers need to have positive feelings by themselves. Teachers' and students' enjoyment in the classroom are positively related to each other (Frenzel, Goetz, Lüdtke, Pekrun, & Sutton, 2009), and teachers' attitudes affect children's mathematical learning (Beilock, Gunderson, Ramirez, & Levine, 2010). Thus, ECEC teacher training focuses not only on mathematical and pedagogical content knowledge but has to deal with prospective teachers' emotions as well. This small-scale study investigates if and how ECEC teacher training at a Norwegian University College changes prospective ECEC teachers' emotions about mathematics.

Theoretical framework

This study focuses on two emotions that we know play an important role in mathematics education: enjoyment and anxiety (Raccanello, Brondino, Moe, Stupnisky, & Lichtenfeld, 2019). The term 'emotion' is difficult to define because it has rather disparate and unspecified meanings (Izard, 2010). Despite all differences, the experts who participated in Izard's study gave definitions that focused on '(a) neural circuits and neurobiological processes, (b) phenomenal experience or feeling, and (c) perceptual-cognitive processes as aspects of emotion' (p. 368). In this present educational study, I do not consider biological aspects. I focus on emotions as phenomenal experiences, feelings like joy and fear. It is possible to 'define functionally discrete emotions like interest, joy, sadness, anger, fear, shame, and guilt' (p. 369). In the context of this study, mathematics-related enjoyment (MJOY) and mathematics anxiety are of special interest. In the following sections, I will describe these concepts in more detail.

Mathematics-related enjoyment

MJOY shows itself in the motive why someone engages in a mathematical activity for his or her own sake and not as a means to gain other rewards (Deci & Ryan, 2000). This is important in Norwegian ECEC institutions because the Norwegian Framework Plan for kindergartens' content and tasks demands that staff encourage children to enjoy mathematics (Norwegian Directorate for Education and Training, 2017). Only a teacher who enjoys mathematics can encourage and facilitate children's enjoyment and will influence their learning, too. Teachers with stronger positive attitudes are more actively establishing good relationships with the children and facilitating the process of learning (Şener, 2015). In addition to mathematical and pedagogical content knowledge, pre-service teachers need positive attitudes towards mathematics to be successful in teaching mathematics (White, Way, Perry, & Southwell, 2005). Enjoyment of teaching mathematics is related to instructional time spent on mathematics (Russo et al., 2020). Teachers' attitudes are correlated with their teaching practices and teachers' MJOY affects the children's enjoyment of mathematics (Stipek, Givvin, Salmon, & MacGyvers, 2001).

Emotions can be analysed as either rapidly changing affective states or relatively stable affective traits (Hannula, 2019, p. 311). In colloquial language, feelings are states that are related to specific situations. We feel enjoyment when we experience a pleasant and enjoyable situation. In a scientific context, enjoyment in a subject is often seen as a stable trait. Strong cognitive dissonance might be needed to change it (Gawronski & Brannon, 2019). In our previous study (Blömeke, Thiel, & Jenßen, 2019), we have shown that prospective ECEC teachers' MJOY is stable across very different situations (with and without an examination) over a period of several weeks. That does not say if it possibly can change throughout a longer period of time such as a year. One aim of the present study is to investigate if ECEC teacher training increases prospective ECEC teachers' enjoyment of mathematics.

Mathematics anxiety

People with *mathematics anxiety* have 'feelings of tension and anxiety that interfere with the manipulation of mathematical problems in a wide variety of ordinary life and academic situations' (Richardson & Suinn, 1972). Anxiety has different aspects (Bessant, 1995): The cognitive aspect covers thoughts about failure; the affective aspect is about emotions like fear and panic; the physiological aspect involves, for example, increased muscle tension; and the behavioural aspect focuses on avoidance. In this education research study, I excluded the physiological consequences of anxiety and focused mainly on cognitive and affective factors. The findings by Jenßen, Dunekacke, Eid, and Blömeke (2015) support that interventions that aim at reducing mathematics anxiety should focus on its stable, cognitive facets.

A meta-study by Carey, Hill, Devine, and Szücs (2016) suggests a bidirectional relationship between mathematics anxiety and performance in mathematics, a vicious cycle: Experienced low performance can cause anxiety and this anxiety can further reduce performance. In a longitudinal study with a representative sample of 3425 German adolescent students (grades 5–9), Pekrun, Lichtenfeld, Marsh, Murayama, and Goetz (2017) found the same reciprocal effects between emotion and achievement. Mathematics-related negative emotions like anxiety were negative predictors of subsequent mathematics achievement, and achievement was a negative predictor for the increase of negative emotions. Our previous study (Thiel & Jenßen, 2018) investigated the second part of this cycle. It revealed the negative impact that prospective ECEC teachers' mathematics anxiety has on their performance in the exam. The present study focuses on the first part. It examines if classroom experiences can change students' mathematics anxiety. Admittedly, I did not measure the students' performance but compared groups with different learning opportunities.

Methodology

The sample

This is a small-scale quantitative longitudinal study with a quasi-experimental design and a convenience sample. An online questionnaire was sent to 392 prospective ECEC teachers from one Norwegian University College at the beginning and the end of the academic year 2017/18. Participation in the study was voluntary and anonymous. In August/September 2017, the response rate was 43%, but in May/June 2018, it was only 16%. I compare three subsamples:

1 **full-time** ECEC teacher students, second year of study, who took a compulsory course in ECEC mathematics education during the project period
2 **part-time** ECEC teacher students, third year of study, with the same compulsory course in ECEC mathematics education during the project period
3 **control** group: part-time ECEC teacher students, second year of study, without any lessons in mathematics during the project period

Independent of their status as a full- or part-time student, all participants in subsamples 1 and 2 received 8–9 ECTS credits in ECEC mathematics education in the project period. ECTS is the European Credit Transfer and Accumulation System. ECTS credits express the volume of learning based on the defined learning outcomes and their associated workload (European Commission, 2015). In addition to lessons in class and self-study, the course included a 35-day practical placement in an ECEC institution and a 5-day practical placement in the first grade of a primary school. The defined mathematical learning outcomes have been that the student:

- has insight into how children develop a comprehensive and flexible mathematical understanding
- has knowledge and skills related to observation, planning, management, and assessment of various play and learning activities and project work with children
- can create a good learning environment in mathematics for all children in kindergarten where exploration, enjoyment in discoveries, and curiosity are given a natural place
- has insight into mathematics education in kindergarten
- can further develop the enjoyment in mathematics that arises in play, everyday situations, and organised activities
- has knowledge of and can reflect on the education in the first year of primary school (Queen Maud University College, 2017)

Measures

In the online questionnaire, I asked for the age of the prospective ECEC teacher and – at the end of the academic year – I added two questions.

1 Do you think your attitude towards mathematics has changed during this academic year? (Answer categories: no, perhaps, yes)
2 If your attitude has changed, what was the main reason for this?

The last question had six answer categories.

1 What I have learnt in the mathematics lessons at the university college
2 What I have learnt in general lessons at the university college
3 What I have experienced during the practical period in primary school
4 What I have experienced during the practical placement in kindergarten
5 Private reasons
6 I don't know

I asked about attitude or stance (Norwegian: *holdning*) and not emotion (Norwegian: *følelse*) because the Norwegian word for emotion is usually used to designate an affective state that appears in a specific situation rather than a permanent trait. After these questions followed the items to measure the prospective ECEC teacher's MJOY and mathematics anxiety.

Mathematics-related enjoyment

To measure prospective ECEC teachers' MJOY, the present study used the same instrument as Blömeke et al. (2019) used in 2015. The participants had to rate the following five items on a 6-point Likert scale from (1) 'strongly disagree' to (6) 'strongly agree':

Mathematics offers the possibility to enjoy discoveries
Mathematics is enjoyable
It is hard to enjoy mathematics
Mathematics leads to enjoyable experiences
Mathematics is boring

For two items (numbers 3 and 5), the scale has been reversed. The reliability of the scale was good (Cronbach's α = .89) in autumn 2017 and excellent (Cronbach's α = .92) in spring 2018. However, Streiner (2003, p. 103) points out that a value 'over .90 most likely indicate unnecessary redundancy rather than a desirable level of internal consistency'. The MJOY score is the average over all five items and can range from 1 to 6.

Mathematics anxiety

To measure prospective ECEC teachers' mathematics anxiety, I used in this study the revised Mathematics Anxiety Scale (MAS-R) (Bai, Wang, Pan, & Frey, 2009). It has 14 items, 6 positive items representing the cognitive aspect and 8 negative items that represent the affective aspect. Participants had to rate the items on a 5-point Likert scale from (1) 'strongly disagree' to (5) 'strongly agree'. The cognitive component MA+ is the sum over all positive items and can range from six to 30. Example items are 'I find maths interesting' and 'Maths relates to my life'. A higher score indicates lower anxiety. The affective component is the sum over all negative items and can range from 8 to 40. Example items are 'Mathematics makes me feel uneasy' and 'I worry about my ability to solve maths problems'. In order to calculate the total score, the positive statements have been scored in reverse so that a high sum score (possibly range from 14 to 70) indicates high anxiety. The same instrument was used by Thiel and Jenßen (2018), and you will find a list of all items there in Table A3. The reliability of the scale was excellent both in autumn 2017 (Cronbach's α = .92) and spring 2018 (Cronbach's α = .91).

The semi-structured interview

After the end of the project period, two prospective ECEC teachers volunteered in a semi-structured interview (Adams, 2015). This was undertaken to gain a deeper insight into the reasons why students changed their emotions about mathematics. The interview guide had the following main questions.

> What did you think about mathematics before you started ECEC teacher education?
> What did you think when you heard that you should have mathematics in ECEC teacher education?
> Have you changed your attitude towards mathematics during ECEC teacher education?
> What do you think about mathematics now?
> Why have you changed your attitude?
> Do you have any suggestions about what we could do better to change students' attitude to mathematics?

Participants' answers were recorded in written note form to protect participants' privacy.

Statistical methods

In the quantitative data analysis, I used as dependent variables MJOY combined with (a) MAS-R and (b) MAS-R's cognitive and affective facets MA+ and MA−. The categorical independent variables were (1) time (with categories Autumn 2017 and Spring 2018), (2) awareness of change (with categories no, perhaps

Table 4.1 Correlations and covariances between all dependent variables

Time	Variable	MJOY	MAS-R	MA+	MA−
Autumn 2017	MJOY	*1.384*	**−0.778**	**0.749**	**−0.661**
(N = 171)	MAS-R	−10.717	*137.112*	**−0.831**	**0.934**
	MA+	4.525	−49.997	*26.389*	**−0.577**
	MA−	−6.193	87.138	−23.615	*63.539*
Spring 2018	MJOY	*1.317*	**−0.769**	**0.787**	**−0.597**
(N = 63)	MAS-R	−9.195	*108.655*	**−0.794**	**0.923**
	MA+	4.230	−38.752	*21.944*	**−0.498**
	MA−	−4.977	69.917	−16.956	*52.853*

Notes: Above the main diagonals in boldface are displayed Pearson correlations. All correlations are significant at the 1% level (two-tailed). On the main diagonals in italics are displayed variances. Below the main diagonals are displayed covariances.

MA+ = cognitive component of MAS-R (positive statements), MA− = affective component of MAS-R (negative statements), MAS-R = mathematics anxiety, MJOY = mathematics-related enjoyment, N = sample size.

and yes), (3) reasons for changed attitudes (with the six categories mentioned in section *Measures*), and (4) subsample (with categories full-time, part-time and control). The dependent variables are strongly correlated (see Table 4.1). The differences between the variance–covariance matrices are not significant (Box's $M = 1.590$, $F(3, 241{,}783.766) = 0.523$, $p = .666$ for MJOY and MAS-R and Box's $M = 3.559$, $F(6, 86{,}432.952) = 0.582$, $p = .745$ for MJOY, MA+, and MA−). Therefore, a multivariate analysis of variance (MANOVA) is the appropriate method to test if the means of the dependent variables are different in the compared groups. I have used IBM SPSS Statistics software version 26.0, a General Linear Model with type III sums of squares, and Tukey's honestly significant difference (Tukey, 1949) as a post hoc test. Since the assumption of balanced group sizes is violated, I have chosen Pillai's Trace V as test statistic because it gives the most robust results under violation of this assumption (Ates, Kaymaz, Kale, & Tekindal, 2019; Olson, 1979).

Findings

Mathematics-related enjoyment

There are significant differences between the beginning and the end of the academic year ($F(3, 230) = 3.728$, $p < .05$; Pillai's trace $V = 0.046$, $\eta_p^2 = 0.046$), but the effect size is small and only related to MJOY. At the beginning of the academic year, the prospective ECEC teachers are ambivalent when it comes to MJOY ($M_{\text{MJOY},1} = 3.75$). At the end of the academic year, they enjoy mathematics significantly more ($M_{\text{MJOY},2} = 4.25$, $\Delta M_{\text{MJOY}} = -.50$, $F(1, 232) = 8.345$, $p < .01$; $\eta_p^2 = 0.035$). This is about the same for four of five items (see Table 4.2; $F(19, 201) = 1.703$, $p < .05$; $V = 0.139$, $\eta_p^2 = 0.139$).

Table 4.2 Item mean and standard deviation of mathematics-related enjoyment by time

	Start autumn 2017			End spring 2018			End spring 2015		
Item	N	Mean (SD)	MD	N	Mean (SD)	MD	N	Mean (SD)	MD
I	168	4.24 (1.25)	4	63	4.65* (1.18)	5	196	4.75 (1.07)	5
2	166	3.56 (1.48)	4	62	4.08* (1.42)	4	196	4.12 (1.27)	4
3	169	3.72 (1.37)	4	63	4.08 (1.40)	4	196	4.20 (1.26)	4
4	168	3.67 (1.27)	4	63	4.30** (1.28)	4	197	4.22 (1.06)	4
5	169	3.58 (1.48)	3	62	4.11* (1.43)	4	196	4.18 (1.32)	4

Notes: This table compares the beginning and the end of the academic year 2017/18 and data from another sample from the end of the academic year 2014/15.

MD = median, N = sample size, SD = standard deviation.

*$p < .05$; **$p < .01$.

Since the response rate at the end of the academic year was much lower than at the beginning, the increase could be due to a bias in the sample. To check if only students who enjoy mathematics participated at the end of the academic year, I compared the data with data from 2015. These data were collected by Blömeke et al. (2019) with a much higher response rate of 64%. Table 4.2 displays the average item scores of MJOY for both time points compared with the data from 2015. The differences between the means from 2018 and 2015 are not significant. Therefore, it is reasonable to assume that the sample is not biased.

Mathematics anxiety

The prospective ECEC teachers' mathematics anxiety is around the centre of the scale. There is no significant difference between the beginning ($M_{MAS-R,1} = 42.99$) and the end ($M_{MAS-R,2} = 41.03$) of the academic year. This is true for the cognitive component ($M_{MA+,1} = 18.60$; $M_{MA+,2} = 19.61$) and for the affective component ($M_{MA-,1} = 25.60$; $M_{MA-,2} = 24.62$). Only one item had in spring 2018 an average rating that was significantly different from autumn 2017. At the end of the academic year, prospective ECEC teachers believed in average slightly more that mathematics relates to their life ($M_{MA05,2} = 3.79$; $\Delta M_{MA05} = -.44$; $F(1, 219) = 5.305 \; p < .05$, $\eta_p^2 = 0.024$). Comparison with data from the 2015 study (Thiel & Jenßen, 2018) shows no significant difference between the different samples, suggesting that the present study's sample is not biased when it comes to mathematics anxiety.

Awareness of change

Of the prospective ECEC teachers who participated in spring 2018, 46% answered that they changed their attitude towards mathematics while 19% answered that they did not change it. Unfortunately, I cannot check if they actually changed because I could not connect the data from the two time points on an individual level. Thus, I can only compare the means of the two groups. Participants who say that they changed their attitude have higher MJOY ($M_{MJOY,yes}$ = 4.51) and lower mathematics anxiety ($M_{MAS-R,yes}$ = 37.6) than participants who answered 'no' ($M_{MJOY,no}$ = 3.55, ΔM_{MJOY} = −.96; $M_{MAS-R,no}$ = 46.2, ΔM_{MAS-R} = 8.6). The participants who answered 'perhaps' have average scores between the other two groups ($M_{MJOY,perhaps}$ = 4.28; $M_{MAS-R,perhaps}$ = 42.4; $M_{MA+,perhaps}$ = 19.9). According to the MANOVA, the differences are not significant on a 5% level ($F(4, 118)$ = 2.278, p = .065; V = 0.143, η_p^2 = 0.072). Even though the effect size is medium (cf. Lakens, 2013; Lenhard & Lenhard, 2016), the observed power of this test is only 0.650. Thus, the problem might be the small sample size. To reach a sufficient test power of at least 0.8 (D'Amico, Neilands, & Zambarano, 2001), we would need at least 78 participants (calculated with the software GLIMMPSE 3.0.0, available at https://v3.glimmpse.samplesizeshop.org/ (Chi, Glueck, & Muller, 2019; Kreidler et al., 2013)).

Reasons for changed attitudes

About half of the participants (56%) answered that the reason for their changed attitude was what they had learnt in the mathematics lessons at the university college while 21% answered that it was what they had experienced during the practical period in an ECEC institution. Only two chose the practical placement in a primary school. One chose personal reasons, and the rest stated that they do not know. Interestingly, there are significant differences between the two main groups ($F(2, 43)$ = 10.073, p < .001; V = 0.319). Participants who chose the lessons have a significantly higher MJOY ($M_{MJOY,lessons}$ = 4.66) and lower mathematics anxiety ($M_{MAS-R,lessons}$ = 36.2) than participants who chose the practical period ($M_{MJOY,practice}$ = 3.7, ΔM_{MJOY} = .94, $F(1, 44)$ = 10.434, p < .01, η_p^2 = 0.192; $M_{MAS-R,practice}$ = 47.5, ΔM_{MAS-R} = −11.3, $F(1, 44)$ = 20.214, p < .001, η_p^2 = 0.315). A partial eta squared (η_p^2) of 0.14 or larger is considered a large effect (Lakens, 2013; Lenhard & Lenhard, 2016). The reason for these differences might be the differing answer patterns in the three subsamples as the next paragraph reveals.

Comparison of the subsamples

Table 4.3 shows the measured MJOY and mathematics anxiety of the three subsamples. Even though the mean scores slightly differ, the differences between the subsamples at the beginning of the academic year are not statistically significant. At the end of the academic year, only the part-time ECEC teacher students who had lessons in mathematics education show higher enjoyment and lower anxiety than the other groups. The differences are significant compared to the beginning of

Table 4.3 Sample size, mean, and standard deviation of mathematics-related enjoyment and mathematics anxiety by subsample and time

Variable	Subsample	Start autumn 2017		End spring 2018		End spring 2015	
		N	Mean (SD)	N	Mean (SD)	N	Mean (SD)
MJOY	Full-time	116	3.67 (1.20)	29	3.92 (.91)	117	3.98 (.90)
	Part-time	36	4.05 (1.09)	21	4.96** (.80)	46	4.94 (.70)
	Control	19	3.69 (1.19)	13	3.83 (1.57)		
MAS-R	Full-time	116	43.0 (12.2)	29	43.3 (9.9)	116	41.5 (10.8)
	Part-time	36	42.3 (10.5)	21	36.1* (9.0)	46	34.6 (9.8)
	Control	19	44.4 (11.8)	13	44.1 (11.3)		
MA+	Full-time	116	18.6 (5.4)	29	18.1 (4.7)	116	19.0 (4.3)
	Part-time	36	19.4 (4.2)	21	22.2* (3.3)	46	21.8 (4.0)
	Control	19	17.9 (5.4)	13	18.7 (5.0)		
MA–	Full-time	116	25.5 (8.2)	29	25.3 (6.6)	116	24.5 (7.6)
	Part-time	36	25.7 (7.4)	21	22.3 (7.1)	46	20.5 (6.8)
	Control	19	26.3 (7.8)	13	26.8 (8.4)		

Notes: This table compares the beginning and the end of the academic year 2017/18 and data from the end of the academic year 2014/15.

MA+ = cognitive component of MAS-R (positive statements), MA– = affective component of MAS-R, MAS-R = mathematics anxiety, MJOY = Mathematics-related enjoyment, N = sample size, SD = standard deviation.

$*p < .05$; $**p < .01$.

the academic year ($F(2, 54) = 6.016$, $p < .01$; $V = 0.182$; $\Delta M_{MJOY} = -.91$, $F(1, 55) = 11.223$, $p < .01$, $\eta_p^2 = 0.169$; $\Delta M_{MAS-R} = 6.2$, $F(1, 55) = 5.155$, $p < .05$, $\eta_p^2 = 0.086$; $\Delta M_{MA+} = -2.8$, $F(1, 55) = 6.855$, $p < .05$, $\eta_p^2 = 0.111$) and between the three subsamples at the end of the academic year ($F(4, 120) = 3.242$, $p < .05$; $V = 0.195$). The differences are significant for both MJOY ($F(2, 60) = 7.259$, $p < .01$; $\eta_p^2 = 0.195$) and mathematics anxiety ($F(2, 60) = 3.910$, $p < .05$; $\eta_p^2 = 0.115$). Pairwise comparisons with Bonferroni adjustment for multiple comparisons show that part-time students have significantly higher MJOY than both full-time students and the control group and significantly lower mathematics anxiety than full-time students. When we split MAS-R in its cognitive and affective component, the MANOVA is still significant ($F(6, 118) = 2.463$, $p < .05$; $V = 0.223$) but only due to a significant difference in the cognitive component ($F(2, 60) = 5.597$, $p < .01$; $\eta_p^2 = 0.157$) between the part-time students and the two other groups. The differences in the affective component of MAS-R are not significant.

Asked about the main reason for changing their attitude towards mathematics, almost all part-time students in the treatment group answered that it was the mathematics lessons at the university college. Only one student chose the practical placement in an ECEC institution as the main reason. On the contrary, many full-time students stated that the seven weeks practical placement in an ECEC institution was the main reason for changing their attitude. This difference is significant (see Table 4.4).

Table 4.4 Reason for changed attitude by subsample

	Lessons	Practice	Total
Full-time	14 (−2.7)	10 (2.7)	24
Part-time	18 (2.7)	1 (−2.7)	19
Total	32	11	43

Notes: Figures are counts (and adjusted standardised residuals in parenthesis).
$\chi^2(1, N = 43) = 7.382, p < .01$.

Interview results

Both participants answered in similar ways. Before they started ECEC teacher education, they thought that mathematics was atrocious and difficult because they did not understand it, needed extra help, and even panicked or cried sometimes. This was not due to one troublesome mathematics classroom experience (Bekdemir, 2010). Rather, it was due to general bad experiences with mathematics content in secondary school and teachers who did not impart mathematics in a comprehensible and engaging way. Asked about their thoughts when they heard that they should have mathematics in ECEC teacher education, one participant answered, 'I thought I just had to get through'. The other thought, 'It can't be that bad. It's probably not about solving equations'.

Both participants said that they changed their attitude towards mathematics during ECEC teacher education because they experienced that it is different from the subject in school. Now they think that ECEC mathematics is funny and important and that it is about understanding, basic skills, inquiry, and not about arithmetic. Asked why they changed their attitudes, both participants agreed that the university college's mathematics lessons played an important role, especially the first lesson that introduced the new subject. One student said that some of the tasks had been too difficult, but both agreed that the tasks should be more related to praxis. Both participants acknowledged that the placement in a kindergarten had a strong impact, too, but not alone. The theory they learnt before the practical placement enabled them to notice and use mathematical learning opportunities in realistic situations and thus see connections between theory and praxis.

The participants suggested using more practical tasks in the mathematics lessons, not merely practical tasks but a good mix of theory and praxis. Furthermore, they wished to reflect on their experiences from the practical placement afterwards together with the mathematics teacher.

Limitations

Two major limitations of this study are the small sample size and that I used a convenience sample. The small sample size results in a low test power, making it difficult to detect medium-sized effects (D'Amico et al., 2001). More problematic is the way I recruited the sample. The MANOVA test assumes a random sample, a condition that is not met here. In a small-scale project without external funding, it is difficult to draw a representative sample. Thus, we have to treat the findings with caution and cannot generalise. Nevertheless shows the study an interesting trend that should be studied in more detail. Especially the very low response rate at the end of the academic year is a problem. There might be a bias if only the most interested students responded. Therefore, I compared my findings with data from an earlier study that used the same instruments at the end of the academic year 2014/15 (Blömeke et al., 2019; Thiel & Jenßen, 2018). In my sample, the MJOY was almost the same, and the mathematics anxiety was only insignificantly higher ($p > .36$ for full-time and $p > .45$ for part-time students) than in 2015 (see Table 4.3). Therefore, I assume that the sample is not biased.

Due to technical problems during the data collection, I could not connect the data from the two time points on an individual level. Therefore, I could only compare group means. It was not possible to analyse intra-individual differences and covariances. This limits the explanatory power of the study as well as the possibilities to check for biases in the sample.

Discussion

The study shows how difficult it is to change prospective ECEC teachers' emotions about mathematics, but it reveals a way that possibly works. Part-time students' MJOY and mathematics anxiety changed positively after one year of study, and they stated that the mathematics lessons were the main reason for the change. In Norway, only 41% of the pedagogical staff in ECEC institutions are trained ECEC teachers (Statistics Norway, 2020). The other employees have other or no formal qualification. The predominant majority of part-time students in ECEC teacher education are working in ECEC institutions and want to become ECEC teachers. Those students already have a lot of practical experience from ECEC but not with teaching mathematics. The lessons at the university college enabled them to apply to their workplace what they have learnt during the theory lessons. Finally, this combination of theory and praxis might have helped them to change their emotions about mathematics. Reflecting on their emotions during practical experiences in light of their theoretical knowledge will be important in this process (Frick, Carl, & Beets, 2010).

The full-time students in my sample attended the same kind of lessons as part-time students, but on average, they did not change their emotions. In difference to the part-time students, the placement in a kindergarten was more important for them. The interviews with two full-time students confirmed the quantitative findings. Both of them explained that the combination of theoretical lessons with practical experiences helped them understand what early childhood mathematics is about, which changed their feelings about the subject. Even though they liked that the practical placement was after the theory lessons, they suggested that it would be even better to get the opportunity to reflect on their experiences in some lessons after the placement. This is a practice that the university college will implement starting from the academic year 2020/21.

Interestingly, only the change in the cognitive component of mathematics anxiety was significant, but not the affective component. This might indicate that only the students' thoughts about mathematics changed, but not their underlying fears. Another explanation can be that it is easier to enhance positive emotions than to overcome mathematics anxiety. In fact, the positively formulated items that measure the cognitive component of the MAS-R do not describe anxiety, but the opposite, appreciation of, interest in and a positive attitude towards mathematics.

Conclusion

Neither practical experience nor theory lessons alone have a measurable effect on prospective ECEC teachers' emotions about mathematics, but the combination of both. Reflecting on experiences from practice based on sound mathematical knowledge for teaching leads to professional competence that will also increase enjoyment and interest in mathematics. This has a distinct practical implication. If ECEC teacher training shall positively affect prospective ECEC teachers' emotions about mathematics, practical and theoretical training should be intertwined. However, because of the limitations of this study, this is just a hypothesis that needs to be confirmed by more research. I suggest a quasi-experimental study comparing four groups: (1) theory lessons followed by a practical placement, (2) a placement followed by theory lessons, (3) intertwined practical and theoretical training, and (4) a control group without mathematics lessons. In addition to studies that focus on inter-individual differences, studies are needed that focus on intra-individual changes.

Acknowledgement

I thank the editors of this volume for their support and valuable suggestions for improvement.

References

Adams, W. C. (2015). Conducting semi-structured interviews. In K. E. Newcomer, H. P. Hatry & J. S. Wholey (Eds.), *Handbook of practical program evaluation* (pp. 492–505). Hoboken, NJ: John Wiley & Sons.

Ates, C., Kaymaz, O., Kale, H. E., & Tekindal, M. A. (2019). Comparison of test statistics of nonnormal and unbalanced samples for multivariate analysis of variance in terms of type-I error rates. *Computational and Mathematical Methods in Medicine, 2019*, 1–8. Retrieved 19 November 2020 from http://downloads.hindawi.com/journals/cmmm/2019/2173638.pdf. https://doi.org/10.1155/2019/2173638

Bai, H., Wang, L., Pan, W., & Frey, M. (2009). Measuring mathematics anxiety: Psychometric analysis of a bidimensional affective scale. *Journal of Instructional Psychology*, *36*(3), 185–193.

Bandura, A. (1977). Self-efficacy: Toward a unifying theory of behavioral change. *Psychological Review*, *84*, 191–215.

Beilock, S. L., Gunderson, E. A., Ramirez, G., & Levine, S. C. (2010). Female teachers' math anxiety affects girls' math achievement. *Proceedings of the National Academy of Sciences of the United States of America*, *107*(5), 1860–1863. https://doi.org/10.1073/pnas.0910967107

Bekdemir, M. (2010). The preservice teachers' mathematics anxiety related to depth of negative experiences in mathematics classroom while they were students. *Educational Studies in Mathematics*, *75*(3), 311–328. https://doi.org/10.1007/s10649-010-9260-7

Bessant, K. C. (1995). Factors associated with types of mathematics anxiety in college students. *Journal for Research in Mathematics Education*, *26*(4), 327–345. https://doi.org/10.2307/749478

Blömeke, S., Thiel, O., & Jenßen, L. (2019). Before, during, and after examination: Development of prospective preschool teachers' mathematics-related enjoyment and self-efficacy. *Scandinavian Journal of Educational Research*, *63*(4), 506–519. https://doi.org/10.1080/00313831.2017.1402368

Carey, E., Hill, F., Devine, A., & Szücs, D. (2016). The chicken or the egg? The direction of the relationship between mathematics anxiety and mathematics performance. *Frontiers in Psychology*, *6*(1987). https://doi.org/10.3389/fpsyg.2015.01987

Carmichael, C., MacDonald, A., & McFarland-Piazza, L. (2014). Predictors of numeracy performance in national testing programs: Insights from the longitudinal study of Australian children. *British Educational Research Journal*, *40*(4), 637–659. https://doi.org/10.1002/berj.3104

Chi, Y.-Y., Glueck, D. H., & Muller, K. E. (2019). Power and sample size for fixed-effects inference in reversible linear mixed models. *The American Statistician*, *73*(4), 350–359. https://doi.org/10.1080/00031305.2017.1415972

D'Amico, E. J., Neilands, T. B., & Zambarano, R. (2001). Power analysis for multivariate and repeated measures designs: A flexible approach using the SPSS MANOVA procedure. *Behavior Research Methods, Instruments, & Computers*, *33*(4), 479–484.

Deci, E. L., & Ryan, R. M. (2000). The "what" and "why" of goal pursuits: Human needs and the self-determination of behavior. *Psychological Inquiry*, *11*(4), 227–268. https://doi.org/10.1207/S15327965PLI1104_01

Duncan, G. J., Dowsett, C. J., Claessens, A., Magnuson, K., & Huston, A. C. (2007). School readiness and later achievement. *Developmental Psychology*, *43*(6), 1428–1446.

Eiser, J. R., Fazio, R. H., Stafford, T., & Prescott, T. J. (2003). Connectionist simulation of attitude learning: Asymmetries in the acquisition of positive and negative evaluations. *Personality and Social Psychology Bulletin*, *29*, 1221–1235.

European Commission (2015). *ECTS users' guide*. Luxembourg: European Union. Retrieved 20 March 2020 from https://op.europa.eu/s/n1Vc

Frenzel, A. C., Goetz, T., Lüdtke, O., Pekrun, R., & Sutton, R. E. (2009). Emotional transmission in the classroom: Exploring the relationship between teacher and student enjoyment. *Journal of Educational Psychology*, *101*(3), 705–716. https://doi.org/10.1037/a0014695

Frick, L., Carl, A., & Beets, P. (2010). Reflection as learning about the self in context: Mentoring as catalyst for reflective development in preservice teachers. *South African Journal of Education*, *30*(3), 421–437.

Gawronski, B., & Brannon, S. M. (2019). What is cognitive consistency, and why does it matter? *Cognitive dissonance: Reexamining a pivotal theory in psychology* (2nd ed., pp. 91–116). Washington, DC: American Psychological Association.

Geary, D. C., Hoard, M. K., Nugent, L., & Bailey, D. H. (2013). Adolescents' functional numeracy is predicted by their school entry number system knowledge. *PLoS ONE, 8*(1), e54651. Retrieved 29 March 2020 from PLOS, website: http://journals.plos.org/plosone/article?id=10.1371/journal.pone.0054651 https://doi.org/10.1371/journal.pone.0054651

Hannula, M. S. (2019). Young learners' mathematics-related affect: A commentary on concepts, methods, and developmental trends. *Educational Studies in Mathematics, 100*(3), 309–316. https://doi.org/10.1007/s10649-018-9865-9

Izard, C. E. (2010). The many meanings/aspects of emotion: Definitions, functions, activation, and regulation. *Emotion Review, 2*(4), 363–370. https://doi.org/10.1177/1754073910374661

Jenßen, L., Dunekacke, S., Eid, M., & Blömeke, S. (2015). The relationship of mathematical competence and mathematics anxiety. *Zeitschrift für Psychologie, 223*(1), 31–38. https://doi.org/10.1027/2151-2604/a000197

Kreidler, S. M., Muller, K. E., Grunwald, G. K., Ringham, B. M., Coker-Dukowitz, Z. T., Sakhadeo, U. R., Barón, A. E., & Glueck, D. H. (2013). GLIMMPSE: Online power computation for linear models with and without a baseline covariate. *Journal of Statistical Software, 54*(10). Retrieved 17 November 2020 from https://www.ncbi.nlm.nih.gov/pmc/articles/PMC3882200/

Lakens, D. (2013). Calculating and reporting effect sizes to facilitate cumulative science: A practical primer for *t*-tests and ANOVAs. *Frontiers in Psychology, 4*, 1–12. Retrieved 16 November 2020 from https://www.frontiersin.org/article/10.3389/fpsyg.2013.00863 https://doi.org/10.3389/fpsyg.2013.00863

Larkin, K., & Jorgensen, R. (2016). 'I hate maths: Why do we need to do maths?' Using iPad video diaries to investigate attitudes and emotions towards mathematics in year 3 and year 6 students. *International Journal of Science and Mathematics Education, 14*(5), 925–944. https://doi.org/10.1007/s10763-015-9621-x

Lenhard, W., & Lenhard, A. (2016). Calculation of effect sizes. Retrieved 17 November 2020 from https://www.psychometrica.de/effect_size.html

Metje, N., Frank, H. L., & Croft, P. (2007). Can't do maths – Understanding students' maths anxiety. *Teaching Mathematics and its Applications: An International Journal of the IMA, 26*(2), 79–88. https://doi.org/10.1093/teamat/hrl023

Norwegian Directorate for Education and Training (2017). *Framework plan for the kindergartens content and tasks.* Oslo: Ministry of Education and Research. Retrieved 29 March 2020 from https://www.udir.no/globalassets/filer/barnehage/rammeplan/framework-plan-for-kindergartens2-2017.pdf

Norwegian Ministry of Education and Research (2015). *Tett på realfag. Nasjonal strategi for realfag i barnehagen og grunnopplæringen (2015–2019) [Close to STEM. National strategy for STEM in kindergarten and primary school (2015–2019)].* Oslo: Kunnskapsdepartementet. Retrieved 4 October 2020 from https://www.regjeringen.no/contentassets/869faa81d1d740d297776740e67e3e65/kd_realfagsstrategi.pdf

OECD. (2019). *PISA 2018 results.* Retrieved 20 March 2020 from www.oecd.org/pisa/publications/pisa-2018- results.htm

Olson, C. L. (1979). Practical considerations in choosing a MANOVA test statistic: A rejoinder to Stevens. *Psychological Bulletin, 86*(6), 1350–1352. https://doi.org/10.1037/0033-2909.86.6.1350

Pekrun, R., Lichtenfeld, S., Marsh, H. W., Murayama, K., & Goetz, T. (2017). Achievement emotions and academic performance: Longitudinal models of reciprocal effects. *Child Development, 88*(5), 1653–1670. https://doi.org/10.1111/cdev.12704

Queen Maud University College. (2017). BHSTM2040 Språk, tekst og matematikk. Retrieved 20 March 2020 from https://studier.dmmh.no/nb/emne/BHSTM2040/272

Raccanello, D., Brondino, M., Moe, A., Stupnisky, R., & Lichtenfeld, S. (2019). Enjoyment, boredom, anxiety in elementary schools in two domains: Relations with achievement. *Journal of Experimental Education, 87*(3), 449–469. https://doi.org/10.1080/00220973.2018.1448747

Remmert, V. R. (2004). What do you need a mathematician for? Martinus Hortensius's "Speech on the dignity and utility of the mathematical sciences" (Amsterdam 1634). *Mathematical Intelligencer, 26*(4), 40–46. https://doi.org/10.1007/bf02985418

Richardson, F. C., & Suinn, R. M. (1972). The mathematics anxiety rating scale: Psychometric data. *Journal of Counseling Psychology, 19*(6), 551–554. https://doi.org/10.1037/h0033456

Russo, J., Bobis, J., Sullivan, P., Downton, A., Livy, S., McCormick, M., & Hughes, S. (2020). Exploring the relationship between teacher enjoyment of mathematics, their attitudes towards student struggle and instructional time amongst early years primary teachers. *Teaching and Teacher Education, 88*, 9. https://doi.org/10.1016/j.tate.2019.102983

Şener, S. (2015). Examining trainee teachers' attitudes towards teaching profession: Çanakkale Onsekiz Mart University Case. *Procedia – Social and Behavioral Sciences, 199*, 571–580. https://doi.org/10.1016/j.sbspro.2015.07.550

Statistics Norway. (2020). Andel barnehagelærere i forhold til grunnbemanning. Retrieved 30 March 2020 from https://www.ssb.no/utdanning/faktaside/barnehager#blokk-3

Stipek, D. J., Givvin, K. B., Salmon, J. M., & MacGyvers, V. L. (2001). Teachers' beliefs and practices related to mathematics instruction. *Teaching and Teacher Education, 17*(2), 213–226. https://doi.org/10.1016/S0742-051X(00)00052-4

Streiner, D. L. (2003). Starting at the beginning: An introduction to coefficient alpha and internal consistency. *Journal of Personality Assessment, 80*(1), 99–103. https://doi.org/10.1207/S15327752JPA8001_18

Thiel, O., & Jenßen, L. (2018). Affective-motivational aspects of early childhood teacher students' knowledge about mathematics. *European Early Childhood Education Research Journal, 26*(4), 512–534. https://doi.org/10.1080/1350293X.2018.1488398

Tukey, J. W. (1949). Comparing individual means in the analysis of variance. *Biometrics, 5*(2), 99–114. https://doi.org/10.2307/3001913

White, A. L., Way, J., Perry, B., & Southwell, B. (2005). Mathematical attitudes, beliefs and achievement in primary preservice mathematics teacher education. *Mathematics Teacher Education and Development, 7*, 33–52.

Chapter 5

A math-avoidant profession?

Review of the current research about
early childhood teachers' mathematics
anxiety and empirical evidence

Lars Jenßen

Introduction

Mathematics is important also from a cultural perspective and the confident handling of mathematical requirements can be understood as a 21st century skill (Goldin, 2014). In our educational and knowledge-based societies, mathematics can therefore also trigger a variety of emotions (ibid.). Early childhood (EC) teachers recognize the importance of mathematics and accordingly express positive attitudes towards mathematics (Benz, 2012; Thiel, 2010). The majority of EC teachers also experience mathematics as emotionally pleasant (Sumpter, 2020). Nevertheless, a significant proportion also reports unpleasant feelings about mathematics (ibid.). It is not surprising that EC teachers regard mathematics as a valuable domain on the one hand and are anxious in situations involving mathematics on the other. According to the control-value theory (Pekrun & Perry, 2014), anxiety can result when people assess a domain as valuable and at the same time assess their own resources to deal with the corresponding requirements as low (e.g. low knowledge in this domain). Indeed, it must be assumed that a significant proportion of EC teachers consider their professional knowledge of mathematics to be low (Blömeke, Jenßen, Grassmann, Dunekacke, & Wedekind, 2017; Noviyanti, 2019). Mathematics anxiety (MA) is one of the most studied unpleasant emotions in mathematics (Dowker, Sarkar, & Looi, 2016) and is also considered relevant for EC teachers. Emotions, understood as affective-motivational dispositions, are a significant facet of teachers' professional competence (Blömeke, Gustafsson, & Shavelson, 2015). It can be assumed that affective-motivational dispositions also show effects on relevant aspects of EC teachers' professional competence (Brown, 2005; Cooke, 2015; Oppermann, Anders, & Hachfeld, 2016). This chapter deals with EC teachers' MA and summarizes the current state of research. An empirical study illustrates whether the profession "EC teacher" can be considered a "math-avoidant" profession.

DOI: 10.4324/9781003172529-5

Phenomenology of mathematics anxiety

MA can be understood as a multifactorial construct that can manifest itself in four different components: affective, cognitive, physiological and behavioral[1] component (Pekrun, Muis, Frenzel, & Goetz, 2018). It is experienced as unpleasant and activating (Pekrun et al., 2018). The affective component of MA can be described as a primary emotional reaction to mathematical demands resulting in fear (Hembree, 1990). Secondarily, helplessness and anger can be experienced. Some authors conceptualize the experience of shame as another primary emotional response (e.g. Wilson, 2017), even though anxiety and shame are two different emotions (Jenßen, Möller, & Roesken-Winter, 2020). Cognitively, thoughts about one's own failure are reported (Hunt, Clark-Carter, & Sheffield, 2014). The thoughts are directed towards the future or evaluate the current situation as challenging or threatening. Even mental blocks are described by people who are afraid of mathematics. Physiologically, above all, tension is described (Hembree, 1990), whereby stronger symptoms such as sweating, a feeling of pressure on the chest or nausea are also possible. The behavioral component of MA describes the avoidance of mathematics-related requirements (Chang & Beilock, 2016), mathematics-related courses during education (Kelly & Tomhave, 1985) or mathematics-related professions (Chipman, Krantz, & Silver, 1992; Huang, Zhang, & Hudson, 2018). Therefore, MA is also conceptualized as the *avoidance of mathematics.* The core feature of MA is that it leads to low mathematical performance, whereby a medium-strong bidirectional effect can be assumed (Carey, Hill, Devine, & Szücs, 2016; Ma, 1999). Women also report higher levels of MA (Sokolowski, Hawes, & Lyons, 2019). MA can occur as an emotional reaction to a specific situation (state) or on a generalized level (trait), as a common emotional tendency of an individual across different situations (Hannula, 2019). In educational research, MA is conceptualized as an emotional disposition that can range from lower MA via medium MA to higher MA (Ashcraft, 2002). However, in clinical psychology, MA can also be conceptualized as a specific phobia with states of intensively experienced panic when working on mathematical tasks (Hembree, 1990).

Early childhood teachers' mathematics anxiety

MA can be considered a poorly researched construct in EC teachers, although many authors consider it a relevant construct for this population. Assumptions about EC teachers' MA are mainly based on results from studies with primary school or secondary school teachers which may not be transferrable to EC teachers. In the following review of the previous studies, a total of 16 studies were considered that explicitly deal with EC teachers' MA (see Table 5.1). Most of the studies come from the United States or Germany and deal with MA of pre-service EC teachers. In the following, the results of the studies are reported separately for pre-service and in-service EC teachers. It has to be noted that MA might have

Table 5.1 Overview of the included studies

Phase of career	Participants	Country	Design	Methodological approach	Type of assessment for mathematics anxiety	Additional variables	Authors
Pre-service	n = 246 K-6 teachers	US	Longitudinal (1 semester)	Mixed methods	Standardized questionnaire (MARS), interviews	None	Gresham (2007)
	n = 156 K-6 teachers	US	Cross-sectional	Mixed methods	Standardized questionnaire (MARS), interviews	Mathematics teacher efficacy	Gresham (2008)
	n = 53 early childhood teachers	Australia	Cross-sectional	Quantitative	Standardized questionnaire (no specific name is given in this paper)	None	Cooke, Cavanagh, Hurst and Sparrow (2011)
	n = 89 early childhood teachers	US	Cross-sectional	Qualitative	Interviews	None	Bates, Latham and Kim (2013)
	n = 30 early childhood teachers (only female)	US	Longitudinal (1 semester)	Mixed methods	Standardized questionnaires (AMAS)	Beliefs, stereotypes in math	Lake and Kelly (2014)
	n = 73 early childhood teachers (results are reported for n = 223 pre-service teachers, where n = 155 were studying primary education)	Australia	Longitudinal (9 weeks)	Mixed methods	Standardized questionnaires (no specific name is given in this paper)	Attitudes towards mathematics, reasons for feeling anxious in mathematics	Boyd, Foster, Smith and Boyd (2014)

(Continued)

Table 5.1 (Continued)

Phase of career	Participants	Country	Design	Methodological approach	Type of assessment for mathematics anxiety	Additional variables	Authors
	n = 354 early childhood teachers	Germany	Longitudinal (3 weeks)	Quantitative	Standardized questionnaire (MAS-R)	Mathematical content knowledge	Jenßen, Dunekacke, Eid and Blömeke (2015)
	n = 12 Pre-K-4 teachers	US	Cross-sectional	Qualitative	Interviews	Mathematics self-efficacy, mathematics teachers' efficacy	Gresham and Burleigh (2018)
	n = 225 early childhood teachers	Norway	Cross-sectional	Quantitative	Standardized questionnaire (MAS-R)	Mathematical knowledge for teaching	Thiel and Jenßen (2018)
	n = 354 early childhood teachers	Germany	Cross-sectional	Quantitative	Standardized questionnaire (MAS-R)	Mathematical content knowledge, mathematics pedagogical content knowledge, math-related perception	Jenßen, Thiel, Dunekacke and Blömeke (2019a)
	n = 392 early childhood teachers	Norway	Longitudinal (1 academic year)	Quantitative	Standardized questionnaire (MAS-R)	Mathematics enjoyment	Thiel (this book)
Pre-service and in-service	n = 100 pre-service early childhood teachers and n = 50 in-service early childhood teachers	Turkey	Cross-sectional	Quantitative	Standardized questionnaire (MAS-R)	Beliefs about mathematics	Aslan (2013)

(Continued)

Phase of career	Participants	Country	Design	Methodological approach	Type of assessment for mathematics anxiety	Additional variables	Authors
	n = 10 K-6 teachers	US	Longitudinal (5 years)	Mixed methods	Standardized questionnaire (MARS), interviews	Mathematics teaching efficacy	Gresham (2018)
	n = 129 early childhood teachers	Germany	Longitudinal (4 years)	Quantitative	Standardized questionnaire (MAS-R)	Mathematics content knowledge, mathematics pedagogical content knowledge, mathematics enjoyment	Jenßen, Eid, Szczesny, Eilerts, and Blömeke (2021)
In-service	n = 20 early childhood teachers and n = 400 children	Turkey	Cross-sectional	Quantitative	Standardized questionnaire (MAS-R)	Beliefs about mathematics, children's mathematics achievement	Aslan, Ogul and Tas (2013)
	n = 48 early childhood teachers and n = 362 children	Germany	Cross-sectional/longitudinal	Quantitative	Standardized questionnaire (MAS-R)	Mathematics content knowledge, children's development of mathematical competence	Jenßen, Hosoya, Jegodtka, Eilerts, Eid and Blömeke (2020)

AMAS = Abbreviated Math Anxiety Scale (Hopko et al., 2003), MARS = Mathematics Anxiety Rating Scale (Richardson & Suinn, 1972), MAS-R = Mathematics Anxiety Rating Scale – Revised (Bai et al., 2009).

different meanings for pre-service EC teachers in comparison to in-service EC teachers. Pre-service teachers might be rather conceptualized as learners than as teachers. Consequently, from the perspective of control-value theory (Pekrun & Perry, 2014), MA might be meaningful for pre-service EC teachers with regard to its effects in achievement and learning situations during teacher education. On the other hand, in-service EC teachers are teaching in practice and MA changes its meaning to a teacher emotion that is rather affected by pedagogical factors (e.g. children's achievement behavior, pedagogical goals of the teacher) than by the domain itself. This assumption is covered by the model of teacher emotions developed by Frenzel (2014). One could argue that the difference is based on the differentiation between MA in achievement situations (pre-service) and MA in pedagogical situations (in-service). Nevertheless, a situation in which the EC teacher's teaching behavior is focused may be perceived additionally as an achievement situation with regard to mathematical content by the EC teacher.

Mathematics anxiety of pre-service early childhood teachers

MA in pre-service EC teachers can manifest itself in different situations: in a learning situation in a mathematics class, in taking a formal mathematics test and in anticipating the teaching of mathematics (Cooke, Cavanagh, Hurst, & Sparrow, 2011). The test situation and the teaching situation have the most challenging character (ibid.). Pre-service EC teachers attribute their MA mainly to their own experience in mathematics (Boyd, Foster, Smith, & Boyd, 2014). The teacher educator's teaching style can be seen as an important factor: too fast progress, poor explanations by the teacher or directive teaching styles can be seen as negative. Self-related beliefs such as "Mathematics is an innate ability" can sustain MA (ibid.). In addition, pre-service EC teachers experience a lower self-confidence in their ability to teach mathematics, which is also due to a lack of knowledge of teaching methods and mathematical knowledge (Bates, Latham, & Kim, 2013). A vicious cycle can be assumed, because poor knowledge in mathematics is also a consequence of the MA. Pre-service EC teachers' MA shows, as expected, negative correlations of medium strength with the professional knowledge in mathematics required for later teaching (Thiel & Jenßen, 2018). The effect of MA seems to be stronger than the effect of mathematics self-efficacy (ibid.). Studies that differentiate professional knowledge into mathematical content knowledge (MCK) and mathematics pedagogical content knowledge (MPCK) showed that MA is mainly associated with MCK (Jenßen, Thiel, Dunekacke, & Blömeke, 2019a; Jenßen, Eid, Szczesny, Eilerts, & Blömeke, 2021). This connection is stable at a trait level under natural conditions and is probably formed mainly by the cognitive component of MA (Jenßen, Dunekacke, Eid, & Blömeke, 2015). The negative relation can be assumed for all mathematical content domains relevant for EC education (number and operations; geometry; measurement and quantity; data and chance), whereby it is stronger for "number and operations" than for "data and chance" (ibid.) For MPCK, the findings are divergent. It can be assumed that MA on MPCK shows only an indirect effect mediated by MCK. This is rather of low-to-medium strength (Jenßen et al., 2019a). The same effect can also be assumed for situation-specific skills such as the professional perception of mathematics-related content in EC educational situations (ibid.). In addition to the cognitive component, MA is also associated with other affective-motivational tendencies, especially mathematics self-efficacy (Gresham & Burleigh, 2018) and mathematics teaching efficacy (Gresham, 2008; Gresham & Burleigh, 2018).

The majority of the studies is concerned with the design of EC teacher education to cope with MA. Lake and Kelly (2014) impressively describe how difficult it is to change MA in a course on early mathematical education (especially didactics). No reduction of MA could be achieved. Boyd et al. (2014) argue that a positive attitude towards mathematics during education can be conveyed if student teachers are made aware of their responsibility for teaching. A key factor appears to be the promotion of conceptual knowledge rather than procedural knowledge and the use of materials to significantly reduce MA (Gresham, 2007). Furthermore, reform-based constructivist methods, peer teaching opportunities and field experience are recommended to address the MA of pre-service EC teachers (Gresham & Burleigh, 2018). Thiel and Jenßen (2018) have found that older student teachers experience less MA. They attributed this result to the fact that these student teachers were able to gain practical experience in EC institutions even before the training. In his current study, Thiel (this book) shows once again that the reflection of practical experience based on professional knowledge in mathematics can help to reduce MA. Practical experience during training thus seems to be a powerful factor in the reduction of MA, but it should be professionally supported within the training.

Mathematics anxiety of in-service early childhood teachers

Surprisingly, a cross-sectional design showed that in-service EC teachers had higher levels of MA than pre-service EC teachers (Aslan, 2013). In fact, however, longitudinal analyses show that MA decreases significantly from training to practice (Gresham, 2018; Jenßen et al., 2021). Nevertheless, the study by Gresham (2018) also shows that MA can still remain high. It has to be noted that explicitly high-math-anxious teachers were interviewed in this study. According to Jenßen et al. (2021), the reduction of MA could be mainly due to the fact that MA probably changes its meaning from anxiety in learning and test situations to anxiety in teaching situations, whereby the construct moves somewhat away from the learning object. This assumption is in line with models of teacher emotions (Frenzel, 2014). An effect in the study of Jenßen et al. (2021) supports this assumption: Existing negative relations to MCK and MPCK during EC teacher education no longer existed in practice. Overall, no clear result is yet evident. Another study with in-service EC teachers in Germany still showed a medium negative association with MCK (Jenßen et al., 2020).

Although it is theoretically assumed (and empirically evident for primary and secondary school teachers, e.g. Hadley and Dorward, 2011; Ramirez, Hooper, Kersting, Ferguson, & Yeager, 2018) that a teacher's emotion can affect the students' achievement (Frenzel, 2014), no study has been able to validate a link between in-service EC teachers' MA and the mathematical competences of children so far (Aslan, Ogul, & Tas, 2013; Jenßen et al., 2020). The study of Aslan et al. (2013) revealed no significant effects of EC teachers' MA on children's mathematical competence. Jenßen et al. (2020) also could not prove a significant effect of EC teachers' MA on children's competence development in mathematics even under control of EC teachers' MCK. However, the authors point out methodological difficulties in adequately modeling the conditions of EC institutions (e.g. several EC teachers are responsible for one child).

Limitations of the existing studies

The studies included in the review vary in sample size and composition. In some studies, only women were examined. The educational level also differs between post-secondary level (vocational schools) and tertiary level (university) due to the international variations in the types of education (Gasteiger, Brunner, & Chen, 2020). In some studies, EC teachers and elementary teachers were examined together, as they are trained together in some countries or the teaching qualification can extend beyond kindergarten to primary school (see Gresham & Burleigh, 2018). This limits the validity of the results of international comparisons. Self-reporting assessments were used in all studies (interviews and questionnaires). Observation methods or physiological methods as objective assessments have not been used so far. The majority of studies focus only on pre-service EC teachers. Only very few studies follow an intervention design. Only in very few studies, it was controlled for cognitive variables (e.g. professional knowledge in mathematics).

Desiderata

The reported studies show that MA can be seen as a relevant variable when examining pre-service as well as in-service teachers' professional competence. Studies showed that MA shows effects on the career choice of math-related professions (e.g. Chipman et al., 1992; Huang et al., 2018), in the sense, that individuals with higher MA tended to avoid these professions. Career choice is a very complex process that is primarily conceptualized as a multifactorial structure of cognitive and social but also of emotional factors (Watt & Richardson, 2007; Wigfield & Eccles, 2000). For prospective teachers, intrinsic and extrinsic motivations are often examined (Watt et al., 2012), and the pedagogical facet of teachers' intrinsic motivation is usually emphasized (Laschke & Blömeke, 2016). When pedagogical professionals are trained as generalists, as this is often the case for primary school teachers and EC teachers, it must be assumed that interest in mathematics is not the primary consideration when choosing this profession.

Additionally, the profession "EC teacher" might be rather associated with pedagogical interests, especially in countries, such as Germany, which show a more social-pedagogical orientation than a school-orientated one. Consequently, one could argue that the profession "EC teacher" is not subjectively associated with mathematics. This in turn would lead to the effect that individuals with higher MA tend to choose this profession for their prospective work. Until now, no study investigated whether MA is a factor that affects the choice of becoming an EC teacher. Or in other words, whether the profession "EC teacher" can be seen as "a math-avoidant profession".

The current study

Background and research question

The current study is part of the *KomMa* project.[2] The main aim of the project was to assess pre-service EC teachers' professional competence in the field of mathematics during their training from beginning to end. In Germany, EC teachers are usually trained at vocational schools as "general educators" for ages 0–18 years (e.g. working later on in the youth welfare system, children's home or preschool) (Gasteiger et al., 2020). The training covers courses in general pedagogy but also includes specific courses in EC education. The number of courses taught specifically related to mathematics within the early years can be seen as very diverse in the vocational schools across all federal states in Germany (Blömeke et al., 2017). Pre-service general educators have to choose at the end of their training, whether they want to work as an EC teacher or whether they want to work in another profession as an educator outside of formal educational institutions (e.g. youth welfare system). In comparison to other countries, they are not allowed to choose the profession "primary school teacher". Consequently, one could say that becoming an EC teacher is the most thinkable formal profession with relations to mathematics for the pre-service educators.

The research question of the present study is whether MA shows effects on pre-service educators' prospective choice during their last year of training to work as an EC teacher. In light of empirical findings and theoretical assumptions, a positive effect of MA on the choice to work as an EC teacher is hypothesized.

Participants, measures and data analysis

To answer the research question, $n = 774$ pre-service educators at the end of their training were asked whether they want to work prospectively as a teacher in EC institutions. Answers could range from "1 = in no case" to "4 = in any case" and represented the variable CHOICE. The present sample covers only those participants of the full *KomMa* sample who were at the end of their training and who were trained at vocational schools (cf. Blömeke et al., 2017). Participants' average age was $M = 23.7$ years (SD = 5.2) and the majority was female (83%).

MA was assessed as an emotional disposition by a standardized questionnaire which was used in PISA 2003 to assess students' MA (Lee, 2009). The adapted questionnaire contains four statements (see Appendix A) which have to be evaluated on a four-point Likert scale ranging from "1 = totally disagree" to "4 = totally agree". Higher values represent a higher level of MA. The questionnaire covers affective (feelings of helplessness), cognitive (worrying) and physiological (nervousness) components of MA. The adapted version shows satisfactory reliability within the current application (Cronbach's alpha = .89) and can be seen as internationally valid (Lee, 2009).

To control for cognitive variables such as professional knowledge, pre-service EC teachers' MCK and MPCK were included in the model to be tested. Educators' MCK and MPCK were tested with the KomMa-MCK-Test and the KomMa-MPCK-Test (Blömeke et al., 2017). Both tests assess specific professional knowledge of EC teachers in Germany, which is gained during their training (Blömeke et al., 2017). The KomMa-MCK-Test contains 24 items covering all relevant mathematical domains for mathematics in the early years (numbers and operations, geometry, quantity and relation, data and chance) and can be assumed as reliable (Rel. = .88). The KomMa-MPCK-Test contains 28 items covering EC teachers' professional knowledge in the domains "diagnosing children's mathematical development" and "designing informal learning environments". The reliability of the test is good (Rel. = .87).

Data were analyzed by applying structural equation modeling. MA was represented by four indicators while MCK and MPCK were included as the latent ability scores, which were created by applying 2PL models for the whole *KomMa* sample (Blömeke et al., 2017). The variable CHOICE was also included as a categorical manifest variable. Correlations between MA, MCK and MPCK were tested as well as regression coefficients from CHOICE on MA, MCK and MPCK.

Results

The raw scores of each variable are reported in Table 5.2. The MA score (sum over all four items) had a potential range of 4–16, so the theoretically expected mean is 10. The empirical mean seems to be comparably high to the theoretically expected mean. The KomMa-Tests were developed such that scores represent

Table 5.2 Descriptive results

	Mathematics anxiety	Mathematics content knowledge	Mathematics pedagogical content knowledge	Prospective choice of ECEC
Mean (standard deviation)	10.28 (3.31)	48.40 (9.64)	49.84 (9.25)	2.72 (.86)
Empirical range (minimum–maximum)	4–16	20–79	15–73	1–4

a normal distribution, so the standardized mean is $M = 50$ and the standard deviation is SD = 10 for the full sample of the *KomMa* project. Consequently, the scores of MCK and MPCK in this sub-sample were closely related to these values. The manifest variable CHOICE had a potential range from min = 1 to max = 4. Higher values indicated a choice in favor of EC education. The mean of CHOICE in the present study was $M = 2.72$. The majority of pre-service educators (59%) answered they can imagine themselves working in the context of EC education (answer at least 3).

The manifest variances and the correlations between all included variables of the structural equation model are presented in Table 5.3. All variances were significant ($p < .001$). MA was negatively associated with MCK and MPCK. The correlation between MA and MCK was stronger than the correlation between MA and MPCK. MCK and MPCK were positively related to each other. The size of the correlation was moderate. MA as well as MCK were related to the choice to work in EC institutions, but MA was positively related and MCK was negatively related. Both sizes of these relations were small. No significant correlation between MPCK CHOICE was found.

To test the hypothesis, the theoretical assumptions were modeled by applying structural equation modeling. MA as a latent trait was measured by four indicators and all factor loadings were significant ($p < .001$) and substantial. Results are presented in Figure 5.1. Curved lines represent correlations and straight lines represent regressive relationships between two variables. The model fitted the data well ($\chi^2(11) = 11.92$, $p = .3694$, RMSEA = .01 [0.00; 0.04], SRMR = .01, CFI = 1.00). MA showed a significant positive effect on CHOICE ($p < .001$). This effect was small. The higher the level of MA, the lower is the prospective choice of working in the EC context. This effect was independent from mathematics knowledge for working as an EC teacher. Both knowledge facets did not show significant effects on the choice to work in EC institutions.

Table 5.3 Manifest variances (diagonal) and manifest correlations (upper triangular matrix) of the variables included in the model

	Mathematics anxiety	Mathematics content knowledge	Mathematics pedagogical content knowledge	Prospective choice of EC education
Mathematics anxiety	.59	−.39	−.20	.14
Mathematics content knowledge	−	92.67	.43	−.11
Mathematics pedagogical content knowledge	−	−	85.38	n.s.
Prospective choice of EC education	−	−	−	.75

Note: All variances and correlations were significant ($p < .001$), except the correlation between MPCK and CHOICE.

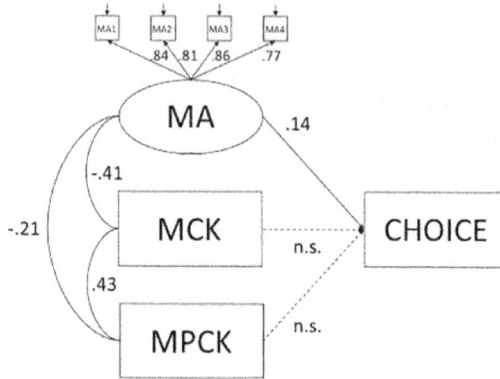

Figure 5.1 Empirical model (standardized solution). *Note:* CHOICE = prospective choice in favor of EC institutions, MA = Mathematics anxiety, MCK = mathematics content knowledge, MPCK = mathematics pedagogical content knowledge.

Discussion

The presented study examined MA as a factor in the career choice of prospective EC teachers.

Previous research showed that MA has negative effects on the choice of a profession in STEM (Chipman et al., 1992; Huang et al., 2018). The present study reveals that mathematics anxious educators are more likely to choose the profession "EC teacher", even if this effect is only of small strength. This could lead to the assumption that EC teacher is indeed *a math-avoidant profession* in Germany. The result of the study is also astonishing as the current study examined the effect of a math-specific construct on general, domain-unspecific, career choice. Thus, the results indicate that the work as an EC teacher is not subjectively seen as a math-specific profession (although mathematics is described as one important domain besides others in the curriculum for EC institutions) – otherwise, it would not make sense that a math-anxious person chooses a math-related career. Nevertheless, the small effect size also indicates that there might be a lot of other aspects that affect the choice in favor of the profession EC teacher (e.g. pedagogical interests).

Unfortunately, the study also shows that the level of mathematics-specific knowledge for EC education (MCK and MPCK) has no effects on career choice: whether educators during their training opt for a career in EC education is independent of whether they have a lot of mathematics content and MPCK.

However, the results of the study must be considered against the background of some limitations and should therefore only be seen as initial empirical evidence. In the present study, only the intentional choice to work in EC institutions was asked and not the actual choice or the motive behind. At least for Germany, it can be assumed that the work as an EC teacher represents the largest occupational field for trained educators. This means that more participants of the study will work in this area later than they stated in their intention in the study. Longitudinal studies would certainly have been advantageous for the investigation of the research question in order to also be able to causally assess the effects of MA. Nevertheless, the results of the study have to be seen as specific to the German context due to the specifics in EC teacher education (e.g. trained as generalists and, especially, the possibility to choose between the profession *EC teacher* and other professions without a focus on children's education).

Conclusion

According to the state of research, MA can be considered a meaningful phenomenon among EC teachers. One could argue that MA in the EC context might be more relevant compared to the school context, because in the less structured EC education context, the avoidance of mathematical situations is much more likely. This may hinder the development of children's competence in mathematics by EC teachers. However, no effects of EC teachers' MA on children's mathematical competence were found in the reviewed studies. Nevertheless, EC teachers are usually trained as generalists, not only in Germany, and are thus trained only to a limited extent in mathematics with regard to EC education (Blömeke et al., 2017). In addition, this circumstance is reinforced by the fact that the formal barrier to teaching mathematics in EC institutions can be considered very low, which in turn can increase pre-service as well as in-service EC teachers' MA. Consequently, EC teachers have fewer resources to meet mathematical requirements in the sense of the control-value theory (Pekrun & Perry, 2014). More importantly, however, the experience of MA should not be seen as competing with the attitude of EC teachers that mathematics is a valuable domain and can therefore be perceived as positive. With regard to control-value theory, this is rather to be understood as an explanation for the experience of MA (Pekrun & Perry, 2014). In educational research on EC teachers' competence, MA is not conceptualized as a pathological phenomenon, but as a relevant disposition in which EC teachers can differ. Some will certainly experience MA in a significant way. The extent to which this occurs, however, might also depend on the way MCK is trained throughout EC teacher education. The more school-based MCK is conceptualized, the higher the level of MA can be, since teachers' MA is also essentially acquired through negative experiences during their own schooling time (Bekdemir, 2010). Which MCK is needed for the work as an EC teacher, however, must still be considered as something to be examined (Jenßen, Dunekacke, Gustafsson, & Blömeke, 2019b). In any case, it must be stated that based on the overall research body about MA, very little research on MA exists among EC teachers.

In the light of the current state of research, the following aspects should be explored in the future:

- Research on effective training programs, which connect mathematics content with reflection of the individual emotional tendencies towards mathematics
- Relations between MA and math-related epistemic beliefs
- Effects of MA on (pre-service) EC teachers' professional performance (e.g. different teaching styles)
- Research on in-service EC teachers' MA and its effects on their professional development
- MA as explanatory factor for possible discrepancies between higher levels of professional knowledge in mathematics and lower levels in professional performance in math-related activities
- Research with focus on EC teachers' mathematics *teaching* anxiety

Notes

1   Some authors label the physiological facet as "somatic" (e.g. Cooke et al., 2011) and the behavioral facet as "motivational". Besides these facets, the expressive facet is also sometimes characterized.
2   *KomMa* was a collaboration of Alice Salomon University of Applied Sciences and Humboldt-Universität zu Berlin from 2012 to 2015, which was funded by the German Federal Ministry of Education (FKZ: 01PK11002A).

References

Ashcraft, M. H. (2002). Math anxiety: Personal, educational, and cognitive consequences. *Current Directions in Psychological Science, 11*(5), 181–185.

Aslan, D. (2013). A comparison of pre- and in-service preschool teachers' mathematical anxiety and beliefs about mathematics for young children. *Academic Research International, 4*(2), 225–230. Retrieved from http://www.savap.org.pk/journals/ARInt./Vol.4(2)/2013(4.2-22).pdf

Aslan, D., Ogul, İ. G., & Tas, I. (2013). The impacts of preschool teachers' mathematics anxiety and beliefs on children's mathematics achievement. *International Journal of Humanities and Social Science Invention, 2*(7), 45–49.

Bai, H., Wang, L., Pan, W., & Frey, M. (2009). Measuring mathematics anxiety: Psychometric analysis of a bidimensional affective scale. *Journal of Instructional Psychology, 36*(3), 189–193.

Bates, A. B., Latham, N. I., & Kim, J. (2013). Do I have to teach math? Early childhood pre-service teachers' fears of teaching mathematics. *IUMPST: The Journal, 5*(August), 1–10.

Bekdemir, M. (2010). The pre-service teachers' mathematics anxiety related to depth of negative experiences in mathematics classroom while they were students. *Educational Studies in Mathematics, 75*(3), 311–328. https://doi.org/10.1007/s10649-010-9260-7

Benz, C. (2012). Maths is not dangerous: Attitudes of people working in German kindergarten about mathematics in kindergarten. *European Early Childhood Education Research Journal*, *20*(2), 249–261. https://doi.org/10.1080/1350293X.2012.681131

Blömeke, S., Gustafsson, J.-E., & Shavelson, R. J. (2015). Competence viewed as a continuum. *Zeitschrift Für Psychologie*, *223*(1), 3–13. https://doi.org/10.1027/2151-2604/a000194

Blömeke, S., Jenßen, L., Grassmann, M., Dunekacke, S., & Wedekind, H. (2017). Process mediates structure: The relation between preschool teacher's education and preschool teachers' knowledge. *Journal of Educational Psychology*, *109*(3), 338–354. https://doi.org/10.1037/edu0000147

Boyd, W., Foster, A., Smith, J., & Boyd, W. E. (2014). Feeling good about teaching mathematics: Addressing anxiety amongst pre-service teachers. *Creative Education*, *05*(04), 207–217. https://doi.org/10.4236/ce.2014.54030

Brown, E. T. (2005). The influence of teachers' efficacy and beliefs regarding mathematics instruction in the early childhood classroom. *Journal of Early Childhood Teacher Education*, *26*(3), 239–257. https://doi.org/10.1080/10901020500369811

Carey, E., Hill, F., Devine, A., & Szücs, D. (2016). The chicken or the egg? The direction of the relationship between mathematics anxiety and mathematics performance. *Frontiers in Psychology*, *6*, 1–6. https://doi.org/10.3389/fpsyg.2015.01987

Chang, H., & Beilock, S. L. (2016). The math anxiety-math performance link and its relation to individual and environmental factors: A review of current behavioral and psychophysiological research. *Current Opinion in Behavioral Sciences*, *10*, 33–38. https://doi.org/10.1016/j.cobeha.2016.04.011

Chipman, S. F., Krantz, D. H., & Silver, R. (1992). Mathematics anxiety and science careers among able college-women. *Psychological Science*, *3*(5), 292–295. https://doi.org/10.1111/j.1467-9280.1992.tb00675.x

Cooke, A. (2015). Considering pre-service teacher disposition towards mathematics. *Mathematics Teacher Education and Development*, *171*, 1–11.

Cooke, A., Cavanagh, R., Hurst, C., & Sparrow, L. (2011). Situational effects of mathematics anxiety in pre-service teacher education. In *Paper presented at the AARE Annual Conference, Hobart 2011* (pp. 1–14).

Dowker, A., Sarkar, A., & Looi, C. Y. (2016). Mathematics anxiety: What have we learned in 60 years? *Frontiers in Psychology*, *7*(Apr.). https://doi.org/10.3389/fpsyg.2016.00508

Frenzel, A. C. (2014). Teacher emotions. In R. Pekrun & L. Linnenbrink-Garcia (Eds.), *International handbook of emotions in education* (pp. 494–519). New York, NY: Routledge. https://doi.org/10.1080/02667363.2014.994350

Gasteiger, H., Brunner, E., & Chen, C.-S. (2020). Basic conditions of early mathematics education: A comparison between Germany, Taiwan and Switzerland. *International Journal of Science and Mathematics Education*, January, 1–17.

Goldin, G. A. (2014). Perspectives on emotion in mathematical engagement, learning, and problem solving. In R. Pekrun & L. Linnenbrink-Garcia (Eds.), *International handbook of emotions in education* (pp. 391–414). New York, NY: Routledge. https://doi.org/10.1080/02667363.2014.994350

Gresham, G. (2007). A study of mathematics anxiety in pre-service teachers. *Early Childhood Education Journal*, *35*(2), 181–188. https://doi.org/10.1007/s10643-007-0174-7

Gresham, G. (2008). Mathematics anxiety and mathematics teacher efficacy in elementary pre-service teachers. *Teaching Education*, *19*(3), 171–184. https://doi.org/10.1080/10476210802250133

Gresham, G. (2018). Preservice to inservice: Does mathematics anxiety change with teaching experience? *Journal of Teacher Education, 69*(1), 90–107. https://doi.org/10.1177/0022487117702580

Gresham, G., & Burleigh, C. (2018). Exploring early childhood preservice teachers' mathematics anxiety and mathematics efficacy beliefs. *Teaching Education*, 1–25. https://doi.org/10.1080/10476210.2018.1466875

Hadley, K. M., & Dorward, J. (2011). Investigating the relationship between elementary teacher mathematics anxiety, mathematics instructional practices, and student mathematics achievement. *Journal of Curriculum and Instruction, 5*(2), 27–44. https://doi.org/10.3776/joci.2011.v5n2p27-44

Hannula, M. S. (2019). Young learners' mathematics-related affect: A commentary on concepts, methods, and developmental trends. *Educational Studies in Mathematics, 100*(3), 309–316. https://doi.org/10.1007/s10649-018-9865-9

Hembree, R. (1990). The nature, effects, and relief of mathematics anxiety. *Journal for Research in Mathematics Education, 21*(1), 33–46.

Hopko, D. R., Mahadevan, R., Bare, R. L., & Hunt, M. K. (2003). The abbreviated math anxiety scale (AMAS) construction, validity, and reliability. *Assessment, 10*(2), 178. https://doi.org/10.1177/1073191103252351

Huang, X., Zhang, J., & Hudson, L. (2018). Impact of math self-efficacy, math anxiety, and growth mindset on math and science career interest for middle school students: The gender moderating effect. *European Journal of Psychology of Education*, September (published online), 1–20. https://doi.org/10.1007/s10212-018-0403-z

Hunt, T. E., Clark-Carter, D., & Sheffield, D. (2014). Math anxiety, intrusive thoughts and performance. *Journal of Education, Psychology and Social Sciences, 2*(2), 69–75.

Jenßen, L., Dunekacke, S., Eid, M., & Blömeke, S. (2015). The relationship of mathematical competence and mathematics anxiety: An application of latent state-trait theory. *Zeitschrift für Psychologie, 223*(1), 31–38. https://doi.org/10.1027/2151-2604/a000197

Jenßen, L., Möller, R., & Roesken-Winter, B. (2020). Shame: A significant emotion in preservice primary school teachers' mathematics education. *Paper accepted for the 14th International Congress on Mathematical Education (Shanghai)*. Cancelled due to the Sars-Cov-2-Pandemic.

Jenßen, L., Thiel, O., Dunekacke, S., & Blömeke, S. (2019a). Mathematikangst bei angehenden frühpädagogischen Fachkräften: Bedeutsam für professionelles Wissen und Wahrnehmung von mathematischen Inhalten im Kita-Alltag? *Journal für Mathematik-Didaktik, 41*, 301–327. https://doi.org/10.1007/s13138-019-00151-1

Jenßen, L., Dunekacke, S., Gustafsson, J.-E., & Blömeke, S. (2019b). Intelligence and knowledge: The relationship between preschool teachers' cognitive dispositions in the field of mathematics. *Zeitschrift für Erziehungswissenschaft, 22*, 1313–1332. https://doi.org/10.1007/s11618-019-00911-2

Jenßen, L., Eid, M., Szczesny, M., Eilerts, K., & Blömeke, S. (2021). Development of early childhood teachers' knowledge and emotions in mathematics during transition from teacher training to practice. *Journal of Educational Psychology*. https://doi.org/10.1037/edu0000518

Jenßen, L., Hosoya, G., Jegodtka, A., Eilerts, K., Eid, M., & Blömeke, S. (2020). Effects of early childhood teachers' mathematics anxiety on the development of children's mathematical competencies. In O. Zlatkin-Troitschanskaia, H. A. Pant, M. Toepper, & C. Lautenbach (Eds.), *Student learning in German higher education* (pp. 141–162). Wiesbaden: Springer Fachmedien Wiesbaden. https://doi.org/10.1007/978-3-658-27886-1

Kelly, W. P., & Tomhave, W. K. (1985). A study of math anxiety/math avoidance in pre-service elementary teachers. *The Arithmetic Teacher*, *32*(5), 51–53.

Lake, V. E., & Kelly, L. (2014). Female preservice teachers and mathematics: Anxiety, beliefs, and stereotypes. *Journal of Early Childhood Teacher Education*, *35*(3), 262–275. https://doi.org/10.1080/10901027.2014.936071

Laschke, C., & Blömeke, S. (2016). Measurement of job motivation in TEDS-M: Testing for invariance across countries and cultures. *Large-Scale Assessments in Education*, *4*(1). https://doi.org/10.1186/s40536-016-0031-5

Lee, J. (2009). Universals and specifics of math self-concept, math self-efficacy, and math anxiety across 41 PISA 2003 participating countries. *Learning and Individual Differences*, *19*(3), 355–365. https://doi.org/10.1016/j.lindif.2008.10.009

Ma, X. (1999). A meta-analysis of the relationship between anxiety toward mathematics and achievement in mathematics. *Journal for Research in Mathematics Education*, *30*(5), 520–540.

Noviyanti, M. (2019). Teachers' belief in mathematics teaching: A case study of early childhood education teachers. *Journal of Physics: Conference Series*, *1315*(1). https://doi.org/10.1088/1742-6596/1315/1/012010

Oppermann, E., Anders, Y., & Hachfeld, A. (2016). The influence of preschool teachers' content knowledge and mathematical ability beliefs on their sensitivity to mathematics in children's play. *Teaching and Teacher Education*, *58*, 174–184. https://doi.org/10.1016/J.TATE.2016.05.004

Pekrun, R., & Perry, R. P. (2014). Control-value theory of achievement emotions. In R. Pekrun & L. Linnenbrink-Garcia (Eds.), *International handbook of emotions in education* (pp. 120–141). New York, NY: Routledge. https://doi.org/10.1080/02667363.2014.994350

Pekrun, R., Muis, K. R., Frenzel, A. C., & Goetz, T. (2018). *Emotions at school*. New York, NY: Routledge. https://doi.org/10.4324/9781315187822

Ramirez, G., Hooper, S. Y., Kersting, N. B., Ferguson, R., & Yeager, D. (2018). Teacher math anxiety relates to adolescent students' math achievement. *AERA Open*, *4*(1), 1–13. https://doi.org/10.1177/2332858418756052

Richardson, F. C., & Suinn, R. M. (1972). The mathematics anxiety rating scale: Psychometric data. *Journal of Counseling Psychology*, *19*(6), 551–554. https://doi.org/10.1037/h0033456

Sokolowski, H. M., Hawes, Z., & Lyons, I. M. (2019). What explains sex differences in math anxiety? A closer look at the role of spatial processing. *Cognition*, *182*, 193–212. https://doi.org/10.1017/CBO9781107415324.004

Sumpter, L. (2020). Preschool educators' emotional directions towards mathematics. *International Journal of Science and Mathematics Education*, *18*, 1169–1184. https://doi.org/10.1007/s10763-019-10015-2

Thiel, O. (2010). Teachers' attitudes towards mathematics in early childhood education. *European Early Childhood Education Research Journal*, *18*(1), 105–115. https://doi.org/10.1080/13502930903520090

Thiel, O., & Jenßen, L. (2018). Affective-motivational aspects of early childhood teacher students' knowledge about mathematics. *European Early Childhood Education Research Journal*, *26*(4), 512–534. https://doi.org/10.1080/1350293X.2018.1488398

Watt, H. M. G., & Richardson, P. W. (2007). Motivational factors influencing teaching as a career choice: Development and validation of the FIT-choice scale. *Journal of Experimental Education*, *75*(3), 167–202. https://doi.org/10.3200/JEXE.75.3.167-202

Watt, H. M. G., Richardson, P. W., Klusmann, U., Kunter, M., Beyer, B., Trautwein, U., & Baumert, J. (2012). Motivations for choosing teaching as a career: An international comparison using the FIT-Choice scale. *Teaching and Teacher Education, 28*(6), 791–805. https://doi.org/10.1016/j.tate.2012.03.003

Wigfield, A., & Eccles, J. S. (2000). Expectancy-value theory of achievement motivation. *Contemporary Educational Psychology, 25*(1), 68–81. https://doi.org/10.1006/ceps.1999.1015

Wilson, S. (2017). Maths anxiety: The nature and consequences of shame in mathematics classrooms. *40 years on: We are still learning! Proceedings of the 40th annual conference of the Mathematics Education Research Group of Australasia* (pp. 562–568).

Appendix A

Items of the questionnaire covering mathematics anxiety (adapted from Lee, 2009)

1   I get very nervous when I have to work on a mathematical task.
2   I worry that I will get wrong results when solving mathematical tasks.
3   I feel helpless when doing a mathematics problem.
4   I worry that it will be difficult for me when working on a mathematical task.

Chapter 6

Kindergarten educators' affective-motivational dispositions

Examining enthusiasm for fostering mathematics in kindergarten

Franziska Vogt, Miriam Leuchter, Simone Dunekacke, Aiso Heinze, Anke Lindmeier, Susanne Kuratli Geeler, Anuschka Meier, Selma Seemann, Andrea Wullschleger, Elisabeth Moser Opitz

Introduction

A teacher's enthusiasm for teaching in general and for the subject they teach in particular is considered to be key for effective teaching (Kunter et al., 2008). Despite the importance of teachers' enthusiasm as well as teacher emotions in general, research into a broad range of teacher emotions is limited, as research often focuses on the specific emotional concerns such as teacher burnout (Frenzel, 2014). Also beyond these, however, teacher emotions are highly relevant. It is widely understood that teacher emotions are "contagious" and thus have an impact on all aspects of teaching and learning. Furthermore, teacher emotions are in a reciprocal relation to other aspects of teaching; for example, teacher enthusiasm might ignite positive emotions amongst students, which in turn lead to more positive emotions of the teacher (Frenzel, 2014). As competence is seen as a continuum ranging from dispositions to situation-specific skills to performance (Blömeke, Gustafsson, & Shavelson, 2015), the role of affective-motivational dispositions warrants more attention.

It is assumed that teacher emotions impact on teaching behaviour, for example "teachers with predominantly positive emotional experiences may be able to effectively utilise a broad range of teaching strategies" (Frenzel, 2014, p. 519). For early childhood mathematics, affective-motivational dispositions regarding the fostering of mathematics might potentially even be more relevant, as educators do not follow a curricular programme but need to seize moments in everyday activities and in play in order to foster mathematics (Link, Vogt, & Hauser, 2017a). Early childhood educators' professional competence is crucial for fostering mathematics in kindergarten (Gasteiger, 2012).

As the following review of research findings illustrates, findings about early educators' affective-motivational dispositions regarding mathematics as a subject and the teaching of mathematics are mixed. Whilst some studies report negative

DOI: 10.4324/9781003172529-6

emotions with regards to the subject of mathematics amongst early educators, other studies found this to be more varied or not the case at all. On the basis of the theoretical considerations outlined, it is proposed that for early childhood educators, the enthusiasm for *fostering* mathematics, thus the affective-motivational dispositions regarding the teaching of the subject rather than the subject itself, is crucial. Research findings on affective-motivational dispositions linked to mathematics in early childhood education are discussed first. Then, the methodology and the results of our study involving 132 early educators in Germany and Switzerland are discussed. The comparison between the two countries is of interest, because the early childhood education system of the two countries differs. Swiss kindergarten is more closely linked to primary school, which might impact on the value placed on mathematics. It is the aim of the study to examine relations between emotions regarding mathematics at school, enthusiasm for mathematics as a subject, importance given to mathematics in kindergarten and educators' self-efficacy as well as the enthusiasm for fostering mathematics in order to add to the understanding of these affective-motivational dispositions as part of early educators' competence.

Theoretical background

First, the review of the research literature focuses on emotions regarding the subject of mathematics. Second, findings on teacher emotions teaching any subject are examined. Third, the enthusiasm for fostering mathematics is discussed. Fourth, a theoretical foundation for exploring the affective-motivational dispositions is provided using the expectancy-value theory. To round off the theoretical background, the similarities and differences of kindergarten in Germany and Switzerland are explained.

Early educators' emotions regarding mathematics

The debate surrounding early educators' emotions regarding mathematics ranges from the concern that many educators have negative emotions in relation to mathematics in comparison to other subjects on the one hand, to the conclusion that emotions are more positive towards the subject than previously expected on the other hand. Anders and Rossbach (2015, p. 310) argue that emotions regarding mathematics develop in school and that "mathematics is one of the subjects that produce the most ambivalent emotional attitudes".

Benz (2012) examined the emotions and beliefs regarding mathematics of 281 kindergarten educators and 308 pre-service kindergarten educators, using questionnaires as a research method. She asked participants to choose adjectives from a selection of 12 adjectives, which could be categorised as positive, neutral, or negative, to express their emotions towards mathematics as a subject. The results reveal that adjectives such as *useful* or *important* are chosen most often, followed by *challenging* and *interesting*. However, a third of the participants also considered mathematics as *confusing*.

Mathematics anxiety as a particular negative emotion could derive from negative school experience, as was found in a study with primary school teachers (Malinsky, Ross, Pannells, & McJunkin, 2006). Early childhood educators might be even more likely to have negative emotions regarding mathematics and "shy away from maths" (Afamasaga-Fuata'i & Sooaemalelagi, 2014, p. 331), as the requirements for mathematics in some countries are lower for early childhood educators than for primary teachers. Perry (2011) identifies that the negative emotions linked to a teacher's own experiences with mathematics at school can translate into a negative attitude to teaching mathematics and a lack of good strategies for mathematics teaching, "a cycle of passing on negative attitudes and strategies for mathematics teaching" (Perry, 2011, p. 8). This cycle needs to be broken.

However, other studies have found that early educators' emotions about mathematics are not as negative as often reported (Chen, McCray, Adams, & Leow, 2014). In a questionnaire study with 110 German kindergarten teachers, a positive attitude towards mathematics was revealed (Thiel, 2010). A study of 221 German kindergarten educators included a questionnaire study with scales on past emotion regarding mathematics at school, current joy in mathematics and the importance given to early mathematics (Anders & Rossbach, 2015). Negative emotions were not as strong as suspected and current joy tended slightly to the positive side. However, the emotions regarding mathematics at school are strongly related to current joy in mathematics. Current joy is also related to the perceived importance of mathematics. In addition, it is interesting to note that past emotions regarding mathematics at school appear not to be related to the importance rating. This result ties in with the study by Benz (2012), who found that educators would state the importance and usefulness of mathematics despite the prevalence of negative emotions towards the subject.

Emotions regarding mathematics appear to correlate with the individual characteristics. Across countries, a gender gap is noted with women revealing more maths anxiety (Marsh et al., 2019). Interestingly, it appears that for early educators, age and work experience are important factors. Older kindergarten teachers were more positive about mathematics than younger ones (Thiel, 2010). In a Norwegian sample of 221 full-time and part-time students for early childhood education, the older students, who largely already got work experience in a kindergarten, showed less mathematics anxiety than their younger, less-experienced, peers (Thiel & Jenßen, 2018). The authors suggest that the practice of fostering mathematics in everyday early childhood practice could help students to overcome mathematics anxieties, which probably derived from their own school experience.

A relevant but less researched field is the interplay between emotions regarding the subject and professional competences. In the above-mentioned study, Anders and Rossbach (2015) also tested the sensitivity of early educators for mathematics in the context of play by using a vignette. They found that the current joy regarding mathematics is related to higher sensitivity for opportunities for fostering mathematics in play. With 354 participating German pre-service kindergarten educators, Jenßen, Thiel, Dunekacke and Blömeke (2020) examined the impact of anxiety and content knowledge as well as pedagogical content knowledge, conceptualised as the "skill to perceive math-related situations during play-based activities" (ibid., p. 301). Mathematics anxiety, measured using questionnaires as well as a video-based assessment, was found to have an indirect effect on the skill to perceive situations during play, which are related to mathematics. This effect was mediated by knowledge. Mathematics anxiety possibly has a large impact: a review of the findings into the direction of causal relationships between mathematics anxiety and mathematics achievement points to a possible vicious circle, whereby anxiety causes lower achievement and lower achievement causes anxiety (Carey, Hill, Devine, & Szücs, 2016).

Teacher's enthusiasm for teaching a subject

Emotions are contagious, as has been stated widely (Frenzel, 2014). Research on teachers' enthusiasm and its effects on students has been able to prove the relationship between teachers' and students' emotions. Becker, Goetz, Morger and Ranellucci (2014) found in a study in grade nine with 149 students in 44 classes that students have an accurate perception of teachers' emotions and that the teachers' emotions affect students in class. The researchers used an experience sampling method across lessons and across four subjects, whereby students were prompted by a device to report on their emotion, their teachers' momentary emotion and the teaching in class. A questionnaire study on positive emotion in mathematics lessons in grade seven and eight also found that teachers' and their students' enthusiasm for the subject are related (Frenzel, Goetz, Lüdtke, Pekrun, & Sutton, 2009). Baumert and Kunter (2006) conceptualise teachers' enthusiasm as part of motivational orientations, together with self-efficacy and self-regulation, enthusiasm representing the emotional aspect within motivational orientations. Blömeke et al. (2015) distinguish between cognitive and affective-motivational dispositions, which would include teacher emotions.

Positive emotions may also strengthen teachers' motivation. If teachers experience positive emotions in the context of teaching a subject, they are motivated to teach the subject again, and to go about teaching the subject with greater engagement, persistence and effectivity (Eccles & Wigfield, 2002 cited in Frenzel, Goetz, & Pekrun, 2008, p. 193). This could lead, for example, to more engagement: teachers might be more motivated to attend professional education, to exchange ideas with colleagues, to read about teaching the subject more widely and to prepare for the subject more intensively (Frenzel et al., 2008).

Kunter et al. (2008) distinguish between enthusiasm for a subject and enthusiasm for teaching a subject and suggest that enthusiasm for teaching a subject is more clearly related to teaching quality than enthusiasm for a subject. In a study of 323 teachers in ninth grade, it was found that teachers' enthusiasm for the subject was related to teachers' self-reported enthusiasm in class but not to students' rating. Whether or not students rated teachers' enthusiasm or teaching quality positively was unrelated to teachers' self-reported enthusiasm for the subject.

Enthusiasm for fostering mathematics

As highlighted in Kunter et al. (2008), enthusiasm for the subject of mathematics is distinct from being enthusiastic about fostering mathematics. Whereas teachers' enthusiasm for mathematics was unrelated, teachers' enthusiasm for teaching mathematics resulted in better ratings of the instructional behaviour by the students. Similarly, a study of 200 prospective US kindergarten teachers shows that emotions towards teaching mathematics were relevant for the quality of teaching, but not the emotions regarding mathematics as a subject per se (Lee, 2005).

In kindergarten, the foundation for mathematics is laid and thus an important and lasting contribution to a child's education is made (Duncan et al., 2007), but the curricular content in itself is basic. A kindergarten teacher might experience positive emotions when supporting a child learning to count and notice the child's progress with excitement, the mere concept of counting objects more likely would not elicit the adult's excitement. Therefore, it could be assumed that the affective-motivational disposition of enthusiasm for fostering mathematics is at the centre within the framework of competence, understood as continuum between dispositions, situation-specific skills and performance as proposed by Blömeke et al. (2015).

As already mentioned, enthusiasm for teaching a subject will increase the motivation to teach the subject further. In the context of early childhood education, where the time spent with a topic is more situational and linked to children's choice as well as to the educator's initiative, positive emotions about fostering mathematics would be even more relevant than in primary school. In early childhood education, the educator's sensitivity to seize the moment in activities and in play is often considered crucial (Rossbach, 2008). A qualitative interview study with German and Swiss educators in kindergarten revealed that educators differ in how purposefully they seek out moments in play and daily activities, which can be used to foster mathematics (Link, Vogt, & Hauser, 2017a).

Expectancy and value

On the basis of expectancy-value theory, proposed for understanding achievement motivation in education in the 1980s by Eccles (Wigfield & Eccles, 2000), the enthusiasm for fostering mathematics in kindergarten would be determined by the confidence of the educator to foster mathematical learning in kindergarten (expectancy) and the subjective value the educator places on mathematics for kindergarten. The expectancy-value theory states that the expectation of succeeding and the subjective value explain a person's achievement motivation, effort, as well as emotions such as interest. The theoretical model implied a multiplicative relation between expectancy and value, which has been modelled in a study with 2508 secondary school students (Trautwein et al., 2012).

For the enthusiasm for fostering mathematics in kindergarten, expectancy is operationalised as self-efficacy regarding fostering mathematics and the value could be expressed in the importance given to mathematics as a subject.

With regards to the expectancy to succeed, the concept of self-efficacy is useful. Self-efficacy has been widely researched in education as well as in psychology. Some aspects relating to early childhood educators are discussed as follows. Brown (2005) found that self-efficacy and importance are correlated; however, they did not predict observed instructional practice. A US study with 67 educators found "that domain-specific self-efficacy was highest for literacy, significantly lower for science, and lowest for math" (Gerde, Pierce, Lee, & Van Egeren, 2018, p. 552). Self-efficacy has unexpected implications. Whilst high self-efficacy is conducive to positive emotions and motivations and might increase the engagement for mathematics in early childhood education, there might also be too much self-efficacy. In relation to mathematics, Thiel and Jenßen (2018) discovered that students who report high self-efficacy in solving mathematical problems might actually perform less well: "In summary, our findings show that MSE [mathematical self-efficacy] has both positive and negative effects on students' achievement. Underestimating one's abilities leads to irrational anxiety, and overestimating one's abilities might cause poor preparation for the exam" (Thiel & Jenßen, 2018, p. 525). For fostering mathematics in kindergarten, we assume that the expectancy of being able to provide such fostering is strengthening educators' affective-motivational dispositions, in particular, the enthusiasm for fostering mathematics.

In order to conceptualise value, research on the value given to mathematics as a subject is discussed. As shown above, Benz (2012) found that mathematics is considered important. Similarly, in the study by Thiel (2010), mathematics as a curricular area for early childhood education remained uncontested. As Benz (2012) explains, the tradition of early childhood education in Germany emphasises less the academic subjects and more social and emotional competences in kindergarten. It is, therefore, important not only to consider the value given to mathematics as a subject but also to compare ratings of the value given to other subjects.

International variation in early childhood education

After having reviewed findings and theories on affective-motivational dispositions regarding mathematics and fostering mathematics in kindergarten, as well as discussing value and expectancy, the structural context of the two countries will be presented, as the affective-motivational dispositions as part of early educators' competence need to be interpreted in the light of the respective education systems and the understanding of early childhood education.

Both countries share an emphasis on play-based pedagogy in kindergarten and, in comparison to other European traditions, emphasise free play, fostering in the situation, and less directive instructional teaching (Rossbach & Grell, 2012). Their pedagogical traditions regarding mathematics in kindergarten are similar (Gasteiger, Brunner, & Chen, 2020).

Although the vast majority of children attend kindergarten before entering first grade of primary school, kindergarten in Germany is not an integral part of the education system. Institutions are led by the state, non-profit organisations such as churches or associations, and by private companies. In contrast, kindergarten in Switzerland is part of the public, state-led education system. Swiss kindergartens are linked to a primary school and are often located in the school buildings themselves or nearby. Kindergarten educators are called "kindergarten teachers". Attending two years of kindergarten is compulsory in Switzerland. In relation to the curricula, in both countries, mathematics is included in the curricula guiding early childhood education. For Switzerland, the kindergarten curriculum is closely aligned to the curriculum of the first and second grade of primary school.

The different kindergarten traditions in the two countries are also noticeable in the work conditions: In Germany, two adults are present and responsible for education and care for a smaller group of children (average adult–child ratio 1:8.3, Destatis, 2020, p. 7). Some of the kindergartens are structured in a so-called open setting, whereby the curricular areas are organised as different rooms providing opportunities for learning activities. Within the team of the educators of a kindergarten, there might be a certain degree of specialisation, in so much, that an educator would be responsible for a particular curricular area and design the learning activities for the whole kindergarten. In contrast, the Swiss kindergarten teachers are for the most part the only adult present with the kindergarten group (average adult–child ratio 1:18, Bundesamt für Statistik, 2020). The Swiss kindergarten teachers are more autonomous in their teaching and solely responsible for designing the learning activities in all areas of the curriculum.

The two countries have different requirements for the qualification of educators in kindergarten. In Germany, early educators mainly train at a professional school at secondary II level, only a small part of the workforce has a bachelor degree. In Switzerland, professional school at secondary II level was the qualification for kindergarten up to approximately the year 2000. From then onwards, kindergarten teachers were required to obtain a bachelor's degree, as for primary school.

Taking these similarities and differences between the two countries into account, which encompass the shared emphasis on a play-based pedagogy, the difference in structure (social pedagogy versus education system, childcare centre versus school), and in the organisation of the work (teamwork versus autonomous teacher), in qualification (professional degree versus higher education bachelor degree), it is therefore of interest whether the affective-motivational dispositions linked to mathematics in early childhood differ between the two countries and depending on qualification.

Research questions

As has been discussed in the review of the literature, affective-motivational dispositions are likely to be relevant for early childhood mathematics, but research on these aspects is still limited. As affective-motivational dispositions should be taken into account when researching educators' professional competence, it remains unclear how these affective-motivational dispositions could be conceptualised and measured. In this study, different scales measuring variables of affective-motivational dispositions are examined. As a theoretical framework, the motivational theory of expectancy and value will be used in order to test relations between several aspects and the crucial concept of enthusiasm for fostering mathematics in kindergarten. To further explore affective-motivational dispositions, a comparison between German and Swiss kindergarten educators is carried out.

The research presented here seeks to describe the affective-motivational dispositions of educators in Germany and Switzerland, as well as seeking to analyse the relations between different aspects.

The main research question guiding this analysis is:

> How are educators' affective-motivational dispositions interrelated, such as the emotions regarding mathematics, the enthusiasm for mathematics as a subject, the value placed on mathematics and the expectancy of fostering mathematics successfully (self-efficacy) and the enthusiasm for fostering mathematics in kindergarten?

Sub-questions are:

- Are there differences between Swiss and German educators or regarding qualification?
- Are emotions about mathematics as a subject best captured with regard to school emotions (retrospectively) or with regard to current enthusiasm for mathematics as a subject?

Methods

The analysis presented here is part of the international research project on the "structure of early childhood educators' domain-specific professional competencies and their effects on the quality of mathematical instructional situations in kindergarten and on children's increase in mathematical competencies" (Moser Opitz, Vogt, Lindmeier, Heinze, & Leuchter, 2015). Within this research project, kindergarten educators filled in a questionnaire which included several scales on aspects of affective-motivational dispositions. First, the sample is described; next, the instrument is discussed.

Sample

For the international study, kindergarten educators were recruited in Northern Germany and German-speaking Switzerland. The questionnaire on educators' beliefs and affective-motivational dispositions was completed by almost all educators (99%) at the beginning of the study. The analysis presented here is based on 64 educators in Germany and 68 educators in Switzerland, in total $N = 132$. As abbreviations for the two countries, D is used for Germany and CH for Switzerland, following the international license plate country code.

For the Swiss sample, 53% were educated at the professional school for kindergarten teachers, the qualification for kindergarten teachers up until the year 2000 and 41% have a Bachelor in pre-primary and primary education, the qualification for kindergarten teachers since 2000. Six per cent had a different qualification.

For the German sample, data about the qualification were missing from eight educators. The majority of educators in the sample, 78%, were qualified kindergarten educators, generally educated at a professional school. Ten per cent were socio-pedagogical assistants and the remaining 12% had a different qualification.

The kindergarten educators had on average 14 years of professional experience ($M = 14.07$; SD = 10.81, $n = 119$). There is a tendency of the German educators in the sample being more experienced ($M_D = 16.33$, $M_{CH} = 12.37$, $t = 1.931$, df = 90.57, $p = .057$).

Scales

In order to measure a variety of affective-motivational dispositions, five scales were selected, adapted and developed. The scales had been used in a tri-national study previously (Link, Vogt, & Hauser, 2017b) and were adapted slightly for the present binational study. In order to measure emotions related to the personal experience with mathematics at school, the scales by Anders and Rossbach (2015, p. 312) regarding positive school emotions ($\alpha = .89$) and negative school emotions ($\alpha = .88$) were consulted. After piloting, one scale on positive emotions on mathematics at school was derived, which was based on four basic emotions, namely fun, pride, anger and anxiety. For enthusiasm for mathematics as a subject, enthusiasm for fostering mathematics in kindergarten and for self-efficacy for fostering mathematics, the same Likert scale ranging from 1 to 6 was used. Enthusiasm for mathematics as a subject was adapted for kindergarten from COACTIV ($\alpha = .82$) (Baumert et al., 2008, p. 96). Table 6.1 provides the items, the response format and the reliability measure.

Questionnaires were administered as paper-and-pencil questionnaires and analysed using the statistics software SPSS. Separately, teachers were asked to provide information about their qualification and work experience.

Table 6.1 Affective-motivational dispositions: scales

Scale	Positive emotions regarding mathematics at school
Items	4
	I was always optimistic that I am able to follow the lessons in mathematics
	I was proud of my achievements in mathematics
	I was annoyed that mathematics was so difficult (reversed)
	In mathematics exams, I felt hopeless (reversed)
Values	1 = does not at all apply; 4 = fully applies
Cronbach's alpha	.887
Scale	**Enthusiasm for mathematics as a subject**
Items	3
	I am myself enthusiastic about mathematics
	I find mathematics exciting and I try to convey this to the children
	I don't like mathematics very much, but I still want the children to have positive experiences with mathematics in kindergarten (reversed)
Values	1 = does not at all apply; 6 = fully applies
Cronbach's alpha	.735
Item	**Importance of mathematics in kindergarten**
Items	1
	Give a personal weighting as to which areas are particularly important to you personally .... Mathematics (other items were language, natural science, music and art, sports)
Values	1 = not at all important; 6 = very important
Scale	**Self-efficacy for fostering mathematics in kindergarten**
Items	4
	I have many ideas how to encourage kindergarten children to learn mathematics
	I feel competent to support children in kindergarten in area of mathematics
	I am very well versed in fostering mathematics in kindergarten
	I feel overcharged to support children in kindergarten in mathematics (reversed)
Values	1 = does not at all apply; 6 = fully applies
Cronbach's alpha	.776
Scale	**Enthusiasm for fostering mathematics in kindergarten**
Items	4
	I have great fun fostering mathematics in kindergarten
	The initiation of mathematical learning processes in the children in my group inspires me
	I find it exciting to watch the children working on mathematical problems
	Fostering mathematics inspires me less than other educational areas of the kindergarten (reversed)
Values	1 = does not at all apply; 6 = fully applies
Cronbach's alpha	.694

Results

In order to describe the affective-motivational dispositions, the results for each of the scales are presented. Second, the correlations between these affective-motivational dispositions, as well as possible differences between German and Swiss educators, are analysed. Third, a multiple regression is performed to conceptualise enthusiasm for fostering mathematics in kindergarten, as it is expected on the basis of the literature, that enthusiasm for fostering mathematics is the most important variable amongst the affective-motivational dispositions for mathematical learning in kindergarten.

Descriptive results on emotions, enthusiasm, value and self-efficacy

The descriptive results of the different scales are discussed first.

> *Positive emotions regarding mathematics at school in retrospect*: Educators were asked in retrospect, how they felt regarding mathematics at school. Their answers were mixed. The scale, which was recoded to express positive emotions regarding mathematics in school, shows an average value only slightly towards the positive ($M = 2.72$; SD = .857, values from 1 to 4). The minimal value of 1, not having had positive emotions at all regarding mathematics at school, has been given by nine educators (7% of the sample), the most positive value of 4, having had mainly positive emotions at school, has been chosen by 12 educators (10%). There is no difference between the two countries, nor regarding the type of qualification (professional college at secondary II level versus higher education at tertiary level).
>
> *Enthusiasm for mathematics as a subject*: The items were formulated in such a way as to referring to the current general emotions regarding mathematics. It shows a large variance: while a third of the educators are less enthusiastic, two-thirds are more enthusiastic ($M = 4.10$, SD = 1.211, values from 1 to 6). Educators with a higher education degree are more enthusiastic about mathematics as a subject than educators qualified at a professional school ($M_{prof} = 3.95$, $SD_{prof} = 1.130$; $M_{HE} = 4.41$, $SD_{HE} = .968$; $t = -2.040$, df = 115, $p = .043$). Swiss educators tend to be more enthusiastic than German educators, a difference, which shows marginal significance and is related to more Swiss educators being qualified at tertiary level ($M_D = 3.93$, $M_{CH} = 4.26$; $t = -1.743$, df = 130, $p = .084$). No correlation was found relating enthusiasm with work experience, educators with more years of working in kindergarten were not more enthusiastic about the subject.
>
> *Importance of mathematics as a curricular content in kindergarten:* Mathematics is considered important, but the importance indicated in comparison other curricular contents is low. All contents are regarded as important or very important. Mathematics, together with natural science,

is considered less important (M_{Math} = 5.08, SD = .647; min = 4 "rather important", max = 6 "very important", values 1–6). Most educators consider mathematics as less important than all other subjects taken together (M_{Maths} = 5.08, $M_{\text{subjects except maths}}$ = 5.39, asympt. Wilcoxon z = –5.595, p < .001). There is a tendency for Swiss educators to consider mathematics more important than German educators (M_{D} =4.98, M_{CH} = 5.18, Mann–Whitney U = 1792.500, p = .095). No difference regarding qualification is found.

Self-efficacy for fostering mathematics in kindergarten: All educators indicate a medium-to-high self-efficacy (M = 4.77, SD = .605, values 1–6). There is a tendency of Swiss educators having higher self-efficacy in comparison to German educators (M_{D} = 4.68; M_{CH} = 4.86, t = –1.172, df = 130, p = .086). There is no difference regarding qualification; however, more work experience is related to higher self-efficacy (r = .211, p = .021).

Enthusiasm for fostering children in mathematics in kindergarten is overall high, all educators are indicating values towards the positive (M = 4.82, SD = .668, min = 3.25, max = 6, values 1–6). There is no difference between Swiss and German educators, nor regarding qualification.

To summarise, educators are valuing all subjects important; however, mathematics is considered less important than other curricular areas. Enthusiasm for fostering mathematics in kindergarten is high, as well as self-efficacy for fostering mathematics. Enthusiasm about mathematics as a subject is mixed, as are emotions referring to mathematics in school. Some of the scales show a slightly more positive affective-motivational disposition amongst the Swiss educators of the sample compared to German educators.

Correlations between affective-motivational dispositions

The affective-motivational dispositions are correlated, as is plausible (Spearman correlations, Table 6.2). Strong correlations can be found between positive emotions regarding mathematics at school and enthusiasm for mathematics as a subject. Also, a factor analysis on the items suggests that these two scales could form one factor. For theoretical reasons however, as emotions regarding mathematics at school specifically refer to the past, whereas enthusiasm for mathematics as a subject is linked to the present, the scales are analysed separately.

Table 6.3 reports the correlations for each of the country sub-samples. Correlations, where German and Swiss educators showed different results, are highlighted in bold. The test used to compare correlations from independent samples is based on Eid, Gollwitzer, and Schmitt (2011).

The correlations between self-efficacy and other variables differ between Switzerland and Germany: self-efficacy is clearly more related to the other affective-motivational dispositions in the German sub-sample. Also for enthusiasm for fostering mathematics, the correlations tend to be higher in the German sub-sample than the Swiss sub-sample.

Table 6.2 Correlations

	Positive emotions regarding mathematics at school	Enthusiasm for mathematics as a subject	Importance of mathematics for kindergarten	Self-efficacy for fostering mathematics in kindergarten
Enthusiasm for mathematics as a subject	.647**			
Importance of mathematics for kindergarten	.204*	.375**		
Self-efficacy for fostering mathematics in kindergarten	.182*	.403**	.401**	
Enthusiasm for fostering mathematics in kindergarten	.269**	.612**	.422**	.530**

$n = 132$; *$p < .05$; **$p < .01$

Table 6.3 Spearman correlations per country

	Positive emotions regarding mathematics at school		Enthusiasm for mathematics as a subject		Importance of mathematics for kindergarten		Self-efficacy for fostering mathematics in kindergarten	
	D	CH	D	CH	D	CH	D	CH
Enthusiasm for mathematics as a subject	.627**	.692**						
Importance of mathematics for kindergarten	.254*	.170	.340**	.384**				
Self-efficacy for fostering mathematics in kindergarten	**.350****	**.063**	**540****	**.237**	**.562****	**.180**		
Enthusiasm for fostering mathematics in kindergarten	**.393****	**.165**	.646**	.552**	**524****	**288***	**.644****	**.363****

Differences between countries; significant in bold and underlined, marginally significant in bold.

CH = Switzerland; D = Germany

$n_D = 63$, $n_{CH} = 68$; *$p < .05$; **$p < .01$

Regression analysis: explaining enthusiasm for fostering mathematics in kindergarten

As has been stated before, enthusiasm for teaching a subject is more important than the enthusiasm for the subject (Kunter et al., 2008; Lee, 2005). We argued that this is especially true for early educators, as the subject content is very basic, the children's learning processes probably being more interesting and stimulating. Therefore, a regression model explaining the enthusiasm for fostering mathematics was tested. It was hypothesised that the enthusiasm for fostering mathematics in kindergarten is predicted by emotions regarding mathematics as a subject, the value given to maths as a subject and the expectancy of self-efficacy.

As for the emotions towards mathematics as a subject, two different scales were examined: positive emotions regarding mathematics as a subject at school in retrospect and enthusiasm for mathematics as a subject. As a sub-question, it is also of interest whether the general emotion about mathematics as a subject or the emotion referring to the experience at school is of more predictive value.

The two measures for emotion regarding mathematics (enthusiasm for mathematics as a subject and positive emotions regarding mathematics at school) were entered first, then the variables for value (importance of mathematics in kindergarten) and for expectancy (self-efficacy for fostering mathematics in kindergarten) were added in order to explain the dependent variable (enthusiasm for fostering mathematics). All conditions were checked carefully. Stepwise regression was calculated first, as the theoretical model assumed an order of influence. However, as results did not differ, the simple entry model is reported.

In this first model, which explains 49.6% of the variance of enthusiasm for fostering mathematics in kindergarten ($R^2 = .496$; $F_{(4,125)} = 32.801$; $p < .001$), enthusiasm for mathematics as a subject remains significant in all steps, whereas positive emotions regarding mathematics at school are no longer significant.

The final model for predicting enthusiasm for fostering mathematics therefore includes enthusiasm for mathematics as a subject, importance of mathematics as a subject in kindergarten and self-efficacy. It explains 48.9% of variance ($R^2 = .489$; $F_{(3,126)} = 42.092$; $p < .001$; $n = 130$). Enthusiasm for mathematics as a subject is the strongest predictor, followed by self-efficacy, and then the importance given to mathematics in kindergarten (Table 6.4).

In this model, the variables "country" (Germany versus Switzerland), "years of work experience" and "qualification" (higher education versus professional education) were entered, but these variables do not have any impact.

As has been found above, some correlations differ significantly between the German and Swiss sub-sample. Therefore, the final multiple regression model was also calculated for each sub-sample separately. The models differ slightly, with the German model having a higher power of prediction: For the model for the German sub-sample, 51.5% variance is explained ($R^2 = .515$; $F_{(3,59)} = 22.987$; $p < .001$; $n = 63$), for the Swiss sub-sample, 42.9% of the variance ($R^2 = .429$; $F_{(3,63)} = 17.544$; $p < .001$; $n = 67$). As Table 6.5 indicates, there are not many differences to be noted, except that for the Swiss educators, the value given to

Table 6.4 Multiple regression analysis explaining enthusiasm for fostering mathematics in kindergarten

	Non-standardised coefficients		Standardised coefficients		Sign
	B	SE	β	T	p
Intercept	1.223	.408		2.997	.003
Enthusiasm for mathematics as a subject	0.250	.044	0.412	5.668	<.001
Value: importance of mathematics in kindergarten	0.197	.072	0.190	2.732	.007
Expectancy: self-efficacy	0.329	.080	0.297	4.111	<.001

Table 6.5 Multiple regression analysis explaining enthusiasm for fostering mathematics in kindergarten for each sub-sample

	Non-standardised coefficients				Standardised coefficients		T		Sign	
	B_D	B_{CH}	SE_D	SE_{CH}	B_D	B_{CH}	T_D	T_{CH}	p_D	p_{CH}
Intercept	.985	1.549	.550	.652			1.792	2.374	.078	.021
Enthusiasm for mathematics as a subject	.213	.292	.067	.062	.352	.481	3.198	4.685	.002	<.001
Value: importance of mathematics in kindergarten	.268	.105	.103	.107	.265	.099	2.609	.981	.011	.330
Expectancy: self-efficacy	.334	.322	.137	.104	.290	.301	2.435	3.104	.018	.003

D = Germany; CH = Switzerland

mathematics as a subject area in kindergarten does not have a significant impact on the enthusiasm for fostering mathematics in kindergarten.

Discussion

Similar to the findings of previous research (Anders & Rossbach, 2015; Benz, 2012), emotions regarding mathematics at school are mixed. There is a considerable part of educators who experienced negative emotions. On the basis of other findings (Thiel & Jenßen, 2018) and the high correlation found in this sample between past and current emotions regarding the subject, it could be assumed that emotions regarding mathematics at school determine enthusiasm for mathematics as a subject to a certain extent. In the regression model, enthusiasm for mathematics is sufficient to explain the impact of the emotions regarding the subject on enthusiasm for fostering mathematics in kindergarten. For future research, it could be sufficient to ask educators generally about enthusiasm for mathematics without having to bring in the retrospective view with the four emotions fun, pride, anxiety and anger. This could be seen as an advantage, as the explicit question about emotions at school risks upsetting participants.

There are some marginally significant differences related to the country: self-efficacy, importance given to mathematics and enthusiasm for fostering mathematics are slightly higher in Switzerland than in Germany. Differences relating to qualification are only found for the enthusiasm for mathematics, with educators with higher education degrees being more enthusiastic about the subject. It could be interpreted that for the affective-motivational dispositions regarding mathematics in kindergarten, the structural differences between the two countries come into play. As kindergarten is defined as part of compulsory schooling in Switzerland, mathematics is given a more distinct weight in the curriculum. This could be reflected in Swiss kindergarten educators rating the importance higher than German educators. Also, the difference in self-efficacy could be linked with the more team oriented approach in Germany, and the role as the sole class teacher of Swiss kindergarten teachers. German educators might share out the responsibility for curricular areas amongst the team and thus might feel less expectations of being particularly competent for mathematics. However, as a limitation, the self-efficacy scale did not focus on the beliefs to handle any problem in the future, as self-efficacy is often measured in other studies (i.e. Schwarzer & Jerusalem, 1999), but was more closely tied to the confidence for fostering mathematics. For future research, the relation between general self-efficacy and self-efficacy for fostering mathematics could be examined. Furthermore, it would be of interest to explore whether the more team oriented or the more isolated work environment of kindergarten teachers does indeed impact these affective-motivational dispositions. Comparative studies including more countries would be needed.

The final model presented above proved to explain 48.9% of the variance in the sample. Enthusiasm for fostering mathematics could be predicted by the emotions for mathematics as a subject, importance given to mathematics as measure of value and self-efficacy as a measure of expectancy. The motivational theory of expectancy and value proves to be a useful theoretical framework for investigating affective-motivational dispositions. The differences between the two countries are slight: the country does not per se explain variance in the multiple regression. However, when analysed separately, as is advised because of the significant differences in the strength of many of the correlations, it shows that for Switzerland, the importance given to mathematics does not have an impact – possibly due to the smaller variance within the Swiss sample. As Swiss educators give higher importance to mathematics than German educators, and as they vary less in their value judgement, this could also be interpreted as being related to kindergarten being part of the school system in Switzerland, but not in Germany.

Limitations of the study presented here are the dependence on self-report measures in the questionnaire. Furthermore, it would be important to link affective-motivational dispositions such as enthusiasm for fostering mathematics in kindergarten with the performance as could be measured through observation. Frenzel et al. (2008) for that reason emphasise the need of seeing emotions on teaching as part of a cycle, leading from the emotion to teaching itself to the learning of the students and to the appraisal of achievement, i.e. the teacher being proud of students' learning. As other findings indicate (Jenßen et al., 2020; Kunter et al., 2008; Lindmeier et al., 2020), also professional knowledge, including both – content knowledge as well as pedagogical knowledge – could also be taken into account in order to understand the ways in which educators' affective-motivational dispositions influence early mathematics education.

Conclusion

Educators' affective-motivational dispositions should be included in any future research into early mathematics in kindergarten as emotions are relevant. Enthusiasm for fostering mathematics is the central variable, but is predicted by enthusiasm for the subject, as well as the expectancy of self-efficacy and the importance given to mathematics. The expectancy-value theory has also proved helpful in this analysis (Marsh et al., 2019) though the model was adapted and the multiplicative relations could not be tested (Trautwein et al., 2012).

The results of the study indicate differences between the two countries which might be linked to characteristics of the education systems. The structure of understanding kindergarten as an integral part of schooling might emphasise the importance of mathematics in kindergarten. However, as the value assigned to mathematics is significantly lower than the values assigned to other curricular areas in both countries, mathematics needs to still be brought into the fore for early childhood education teachers.

The findings underline the importance of affective-motivational dispositions in relation to fostering mathematics. It could be concluded that professional development for increasing competences for fostering mathematics in kindergarten in providing training for situation-specific skills should also address affective-motivational dispositions. Igniting enthusiasm for mathematics in kindergarten and enhancing the expectancy of self-efficacy might lead to a higher commitment for fostering mathematics in early childhood education. To achieve an improvement, initial professional education as well as professional development would benefit from not only ensuring the acquisition of cognitive dispositions and skills but also from an emphasis on affective-motivational dispositions for early mathematics, through strengthening enthusiasm for fostering mathematics in kindergarten.

Acknowledgements

We thank all the participating educators and children in kindergartens in Germany and Switzerland.

The project was funded by a grant of the Swiss National Science Foundation (Grant 156680) and by the German Research Foundation (LI 2616/1-1, HE 4561/8-1/LE 3327/2-1).

Open access of this chapter was funded by the Open Access Publication Fund of the St. Gallen University of Teacher Education (PHSG).

References

Afamasaga-Fuata'i, K., & Sooaemalelagi, L. (2014). Student teachers' mathematics attitudes, authentic investigations and use of metacognitive tools. *Journal of Mathematics Teacher Education, 17*(4), 331–368.

Anders, Y., & Rossbach, H.-G. (2015). Preschool teachers' sensitivity to mathematics in children's play: The influence of math-related school experiences, emotional attitudes and pedagogical beliefs. *Journal of Research in Childhood Education, 29*(3), 305–322. https://doi.org/10.1080/02568543.2015.1040564

Baumert, J., & Kunter, M. (2006). Stichwort: Professionelle Kompetenz von Lehrkräften. *Zeitschrift für Erziehungswissenschaft, 9*(4), 469–520.

Baumert, J., Blum, W., Brunner, M., Dubberke, T., Jordan, A., Klusmann, U., Krauss, S., Kunter, M., Löwen, K., Neubrand, M., & Tsai, Y.-M. (2008). *Professionswissen von Lehrkräften, kognitiv aktivierender Mathematikunterricht und die Entwicklung von mathematischer Kompetenz (COACTIV): Dokumentation der Erhebungsinstrumente.* Berlin: Max-Planck-Institut für Bildungsforschung. https://pure.mpg.de/rest/items/item_2100057/component/file_2197666/content (accessed 2.10.2020).

Becker, E. S., Goetz, T., Morger, V., & Ranellucci, J. (2014). The importance of teachers' emotions and instructional behavior for their students' emotions: An experience sampling analysis. *Teaching and Teacher Education, 43*(1), 15–26. https://doi.org/10.1016/j.tate.2014.05.002

Benz, C. (2012). "Maths is not dangerous"—Attitudes of people working in German kindergarten about mathematics in kindergarten. *European Early Childhood Education Journal, 20*(2), 249–261. https://doi.org/10.1080/1350293X.2012.681131

Blömeke, S., Gustafsson, J.-E., & Shavelson, R. J. (2015). Beyond dichotomies: Competence viewed as a continuum. *Zeitschrift für Psychologie, 223*(1), 3–13. https://doi.org/10.1027/2151-2604/a000194

Brown, E. T. (2005). The influence of teachers' efficacy and beliefs regarding mathematics instruction in the early childhood classroom. *Journal of Early Childhood Teacher Education, 26*(3), 239–257.

Bundesamt für Statistik (2020). *Obligatorische Schule: Betreuungsverhältnis.* Neuchatel: Bundesamt für Statistik. https://www.bfs.admin.ch/bfs/de/home/statistiken/bildung-wissenschaft/bildungsindikatoren/bildungsstufen/obligatorische-schule/betreuungsverhaeltnis.html (accessed 2.10.2020).

Carey, E., Hill, F., Devine, A., & Szücs, D. (2016). The chicken or the egg? The direction of the relationship between mathematics anxiety and mathematics performance. *Frontiers in Psychology, 6*, 1–6. https://doi.org/10.3389/fpsyg.2015.01987

Chen, J.-Q., McCray, J., Adams, M., & Leow, C. (2014). A survey study of early childhood teachers' beliefs and confidence about teaching early math. *Early Childhood Education Journal, 42*(6), 367–377. https://doi.org/10.1007/s10643-013-0619-0

Destatis (2020). *Der Personalschlüssel in Kindertageseinrichtungen Methodische Grundlagen und aktuelle Ergebnisse.* Statistisches Bundesamt. https://www.destatis.de/DE/Themen/Gesellschaft-Umwelt/Soziales/Kindertagesbetreuung/Publikationen/Downloads-Kindertagesbetreuung/kindertageseinrichtungen-personalschluessel-5225409199004.pdf;jsessionid=78AC824099616A81C0A1153574E30596.internet8712?__blob=publicationFile (accessed 30.11.2020).

Duncan, G. J., Dowsett, C. J., Claessens, A., Magnuson, K., Huston, A. C., Klebanov, P., Pagani, L. S., Feinstein, L., Engel, M., Brooks-Gunn, J., Sexton, H., Duckworth, K.,

& Japel, C. (2007). School readiness and later achievement. *Developmental Psychology*, *43*(6), 1428–1446. https://doi-org/10.1037/0012-1649.44.1.217

Eid, M., Gollwitzer, M., & Schmitt, M. (2011). *Statistik und Forschungsmethoden*. Weinheim: Beltz. Test published online: https://www.psychometrica.de/correlation. html (accessed 01.10.2020).

Frenzel, A. C. (2014) Teacher emotions. In E. A. Linnenbrink-Garcia & R. Pekrun (Eds.), *International handbook of emotions in education* (pp. 494–519). New York, NY: Routledge.

Frenzel, A. C., Goetz, T., Lüdtke, O., Pekrun, R., & Sutton, R. E. (2009). Emotional transmission in the classroom: Exploring the relationship between teacher and student enjoyment. *Journal of Educational Psychology*, *101*(3), 705–716. http://dx.doi. org/10.1037/a0014695

Frenzel, A. C., Goetz, T., & Pekrun, R. (2008). Ursachen und Wirkungen von Lehreremotionen: Ein Modell zur reziproken Beeinflussung von Lehrkräften und Klassenmerkmalen. In M. Gläser-Zikuda (Ed.), *Lehrerexpertise: Analyse und bedeutung unterrichtlichen handelns* (pp. 187–209). Münster: Waxmann.

Gasteiger, H. (2012). Fostering early mathematical competencies in natural learning situations: Foundation and challenges of a competence-oriented concept of mathematics education in kindergarten. *Journal für Mathematik-Didaktik*, *33*(2), 181–201. https:// doi.org/10.1007/s13138-012-0042-x

Gasteiger, H., Brunner, E., & Chen, C. S. (2020). Basic conditions of early mathematics education: A comparison between Germany, Taiwan and Switzerland. *International Journal of Science and Mathematics Education*, *19*, 1–17. https://doi.org/10.1007/ s10763-019-10044-x

Gerde, H. K., Pierce, S. J., Lee, K., & Van Egeren, L. A. (2018). Early childhood educators' self-efficacy in science, math, and literacy instruction and science practice in the classroom. *Early Education and Development*, *29*(1), 70–90. https://doi.org/10.1080/ 10409289.2017.1360127

Jenßen, L., Thiel, O., Dunekacke, S., & Blömeke, S. (2020). Mathematikangst bei angehenden frühpädagogischen Fachkräften: Bedeutsam für professionelles Wissen und Wahrnehmung von mathematischen Inhalten im Kita-Alltag? *Journal für Mathematik-Didaktik*, *41*, 301–327. https://doi.org/10.1007/s13138-019-00151-1

Kunter, M., Tsai, Y. M., Klusmann, U., Brunner, M., Krauss, S., & Baumert, J. (2008). Students' and mathematics teachers' perceptions of teacher enthusiasm and instruction. *Learning and Instruction*, *18*(5), 468–482. https://doi.org/10.1016/j.learninstruc.2008.06.008

Lee, J. (2005). Correlations between kindergarten teachers' attitudes toward mathematics and teaching practice. *Journal of Early Childhood Teacher Education*, *25*(2), 173–184. https://doi.org/10.1080/1090102050250210

Lindmeier, A., Seemann, S., Kuratli-Geeler, S., Wullschleger, A., Dunekacke, S., Leuchter, M., Vogt, F., Opitz, E. M., & Heinze, A. (2020). Modelling early childhood teachers' mathematics-specific professional competence and its differential growth through professional development: An aspect of structural validity. *Research in Mathematics Education*, *22*(2), 168–187. https://doi.org/10.1080/14794802.2019.1710558

Link, M., Vogt, F., & Hauser, B. (2017a). Weil durch Zwingen lernen sie es sowieso nicht: Überzeugungen pädagogischer Fachkräfte zum mathematischen Lernen im Kindergarten. In S. Schuler, C. Streit, & G. Wittmann (Hrsg.) *Perspektiven mathematischer Bildung im Übergang vom Kindergarten zur Grundschule* (pp. 255–226). Wiesbaden: Springer Spektrum.

Link, M., Vogt, F., & Hauser, B. (2017b). Überzeugungen von Kindergartenlehrpersonen zur mathematischen Förderung im Kindergarten: Die Schweiz, Deutschland und Österreich im Vergleich. *Beiträge zur Lehrerinnen- und Lehrerbildung, 35*(3), 440–458.

Malinsky, M., Ross, A., Pannells, T., & McJunkin, M. (2006). Math anxiety in pre-service elementary school teachers. *Education, 127*(2), 274–279.

Marsh, H. W., Van Zanden, B., Parker, P. D., Guo, J., Conigrave, J., & Seaton, M. (2019). Young women face disadvantage to enrollment in university STEM coursework regardless of prior achievement and attitudes. *American Educational Research Journal, 56*(5), 1629–1680. https://doi.org/10.3102/0002831218824111

Moser Opitz, E., Vogt, F., Lindmeier, A., Heinze, A., & Leuchter, M. (2015). Struktur fachspezifischer professioneller Kompetenzen von pädagogischen Fachkräften und ihre differenziellen Effekte auf die Qualität von mathematischen Lehr-Lern-Situationen im Kindergarten und den Kompetenzzuwachs von Kindern. http://p3.snf.ch/project-156680# (accessed 2.10.2020).

Perry, C. A. (2011). Motivation and attitude of preservice elementary teachers toward mathematics. *School Science and Mathematics, 111*(1), 2–10. https://doi.org/10.1111/j.1949-8594.2010.00054.x

Rossbach, H.-G. (2008). Was und wie sollen Kinder im Kindergarten lernen? In H.-U. Otto & T. Rauschenbach (Eds.), *Die andere Seite der Bildung* (pp. 123–131). Wiesbaden: VS Verlag für Sozialwissenschaften.

Rossbach, H.-G., & Grell, F. (2012). Vorschulische Einrichtungen. In U. Sandfuchs, W. Melzer, B. Dühlmeier & A. Rausch (Eds.), *Handbuch Erziehung* (pp. 332–337). Bad Heilbrunn: Julius Klinkhardt.

Schwarzer, R., & Jerusalem, M. (Eds.) (1999). *Skalen zur Erfassung von Lehrer- und Schülermerkmalen. Dokumentation der psychometrischen Verfahren im Rahmen der Wissenschaftlichen Begleitung des Modellversuchs Selbstwirksame Schulen.* Berlin: Freie Universität Berlin.

Thiel, O. (2010). Teachers' attitudes towards mathematics in early childhood education. *European Early Childhood Education Research Journal, 18*(1), 105–115. https://doi.org/10.1080/13502930903520090

Thiel, O., & Jenßen, L. (2018). Affective-motivational aspects of early childhood teacher students' knowledge about mathematics. *European Early Childhood Education Research Journal, 26*(4), 512–534. https://doi-org/10.1080/1350293X.2018.1488398

Trautwein, U., Marsh, H. W., Nagengast, B., Lüdtke, O., Nagy, G., & Jonkmann, K. (2012). Probing for the multiplicative term in modern expectancy–value theory: A latent interaction modeling study. *Journal of Educational Psychology, 104*(3), 763–777. https://doi.org/10.1037/a0027470

Wigfield, A., & Eccles, J. S. (2000). Expectancy–value theory of achievement motivation. *Contemporary Educational Psychology, 25*(1), 68–81. https://doi.org/10.1006/ceps.1999.1015

Part II

Situation-Specific Skills

Chapter 7

Early mathematics education

What do pre-service teachers learn?

Simone Dunekacke, Sigrid Blömeke

Introduction

Until recently, early childhood (EC) teacher education was a field in which research was scarce. Consequently, little is known about how pre-service EC teachers acquire professional competence (La Paro, van Schagen, King, & Lippard, 2017). This problem is even more evident when it comes to the acquisition of domain-specific competences in mathematics and other domains (Horm, Hyson, & Winton, 2013; Paprzycki et al., 2017) or to EC teacher education in different types of institutions, such as vocational versus university settings (Oberhuemer, Schreyer, & Neuman, 2010). In this chapter, we aim to provide more insight into German EC teacher education in the vocational track, where 95% of EC teacher training in Germany takes place (Autorengruppe Fachkräftebarometer, 2019). We investigate differences in the acquisition of professional competence between pre-service EC teachers in different years of teacher education with regard to their mathematics pedagogical content knowledge (MPCK) and their situation-specific skills, namely the skill in perceiving math-related situations in EC education and the skill in planning math-related actions.

In both teacher education in general (Blömeke, Gustafsson, & Shavelson, 2015; Depaepe, Verschaffel, & Star, 2020) and EC teacher education (Fröhlich-Gildhoff, Nentwig-Gesemann, & Pietsch, 2011; Gasteiger, Brunner, & Chen, 2020), teacher competence is typically defined as a multi-dimensional construct consisting of dispositions, which in turn encompass cognitive and non-cognitive facets, as well as of situation-specific skills and observable behaviour. The corresponding conceptual state of research is summarized within this volume by Dunekacke, Jegodtka, Koinzer, Eilerts and Jenßen as well as by Brunner. Teacher competence is seen as highly domain-specific and as learnable – during teacher education, for example (Blömeke, Gustafsson et al., 2015; Hartig & Klieme, 2006). In this article, we focus on one facet of pre-service EC teachers' professional knowledge as well as on their situation-specific skills.

MPCK has been identified as an important facet of teachers' professional knowledge, in addition to mathematical content knowledge and general pedagogical knowledge (Ball & Bass, 2009; Shulman, 1986). MPCK is understood

DOI: 10.4324/9781003172529-7

as declarative knowledge (Anderson & Krathwohl, 2001) addressing factual and conceptual aspects of early mathematics education, such as models of domain-specific child development or typical instructional strategies. EC teachers' MPCK has become a growing field of research within the last years (Linder & Simpson, 2018). This research has indicated that it is possible to reliably and validly assess MPCK (Blömeke, Jenßen, Grassmann, Dunekacke, & Wedekind, 2017; Gasteiger, Bruns, Benz, Brunner, & Sprenger, 2020; Lee, 2010). Research linking structural aspects of teacher education to its outcomes has found that opportunities to learn MPCK as well as mathematical content knowledge in German teacher education are rare, but when they do exist, they are positively related to the development of domain-specific knowledge (Blömeke et al., 2017; Kleeberger & Stadler, 2011). Furthermore, research indicates that pre-service EC teachers' professional knowledge differs depending on their year of teacher education and the type of training institution (Blömeke et al., 2017; Torbeyns, Verbruggen, & Depaepe, 2019). Torbeyns and colleagues (2019) showed that pre-service EC teachers in their first year of study have significant lower knowledge than those in their second and third year. Furthermore, they showed that second- and third-year students did not significantly differ in their MPCK.

There is an ongoing discussion about how professional knowledge has an effect on teachers' performance in class and student learning (Blömeke, Gustafsson et al., 2015; Depaepe et al., 2020; Depaepe, Verschaffel, & Kelchtermans, 2013; Kersting, Givvin, Sotelo, & Stigler, 2010; Santagata & Yeh, 2016). Situation-specific skills, namely the skill to perceive learning situations from a domain-specific perspective and to plan adequate actions for such situations, have been discussed as useful for teachers because they may link professional knowledge and performance (Blömeke, Gustafsson et al., 2015; Depaepe et al., 2020). These skills have been identified in theory and research as potential mediators specifically between MPCK and teacher performance (Blömeke, Gustafsson et al., 2015; Depaepe et al., 2020; Dunekacke, Jenßen, Eilerts, & Blömeke, 2016; McCray & Chen, 2012; Meschede, Fiebranz, Möller, & Steffensky, 2017; Santagata & Yeh, 2016).

Applying the construct of situation-specific skills implies taking a more "situated perspective on teachers' knowledge" (Stahnke, Schueler, & Roesken-Winter, 2016). Such skills address the question of what teachers notice in a given teaching situation and how their noticing can be linked to teaching performance and student learning in a domain (van Es & Sherin, 2008). How teachers balance the broad perception of a wide range of classroom aspects and the more in-depth perception of unique, domain-specific aspects is an ongoing discussion (Stahnke et al., 2016). However, researchers agree that both are necessary, but the skills might depend on different kinds of knowledge (Stahnke et al., 2016). This chapter focuses on domain-specific aspects, namely skill in perceiving math-related aspects of teaching situations. This includes structural aspects of the situation, such as the materials used, as well as process-related aspects, such as the children's domain-specific developmental stage (Star & Strickland, 2008; van Es & Sherin, 2008).

Existing research has mostly used different kinds of vignettes to investigate the skill in perceiving math-related situations. Such vignettes are based on short verbal descriptions (Oppermann, Anders, & Hachfeld, 2016), pictures (Lindmeier et al., 2020; Lindmeier, Hepberger, Heinze, & Moser Opitz, 2016; Wittmann, Schuler, & Levin, 2015), or video vignettes (Bruns & Gasteiger, 2019; Bruns, Strahl, & Gasteiger, 2020; Dunekacke et al., 2016; Dunekacke, Jenßen, & Blömeke, 2015). Research on the relationship between mathematical content knowledge and the skill in perceiving math-related situations has not yet come to consistent results. Dunekacke et al. (2015) found a positive relationship between the two constructs among pre-service EC teachers. In contrast, Oppermann et al. (2016) found no significant relationship for in-service teachers but could show that mathematical self-efficacy was a predictor of the skill in perceiving math-related situations. Dunekacke et al. (2016) showed that mathematical content knowledge might be a pre-condition for MPCK, which in turn was related to skill in perceiving math-related situations. Lindmeier and colleagues (2016, 2020) found a correlation between aspects of pedagogical content knowledge and the skill in perceiving math-related situations.

Situation-specific skills also address the question of how perception is linked to interactions within a teaching situation (Stahnke et al., 2016; van Es & Sherin, 2008). In EC settings, this becomes even more important and often takes the form of a situation-specific approach, where children and EC teachers together decide which aspects of a situation to take up and work with. We address this aspect from a domain-specific perspective as well, focusing on the planning of math-related actions as another situation-specific skill. Such planning processes occur on two levels. On the one hand, planning can take place within a situation, which is sometimes also known as decision-making. On the other hand, there are aspects of planning with a longitudinal perspective. The former type of planning is particularly common among EC teachers in Germany, because young children's learning mostly takes place in everyday life and play-based situations rather than via planned activities (Gasteiger & Benz, 2018). Research on planning math-related actions is limited. Dunekacke et al. (2015, 2016) showed a strong correlation between the skill in perceiving math-related actions and the skill in planning math-related actions, but they could not rule out that this result occurred for methodological reasons. Lindmeier et al. (2016, 2020) found significant correlations between knowledge, perception and planning action among in-service teachers.

To the best of our knowledge, there is no research on differences in the skill in perceiving math-related situations and the skill in planning math-related actions among pre-service EC teachers in different years of teacher education.

Early childhood teacher education in Germany

Vocational-track EC teacher education in Germany takes place at specialized schools (*Fachschulen/Fachakademien*) and is of varying duration depending on the federal state (Oberhuemer et al., 2010). Until the early 2000s, this form of teacher education focused on providing pre-service teachers with the competence to support the social-pedagogical development of children (Oberhuemer et al., 2010). However, a discussion about a potentially low cognitive quality of German EC education developed at this time, especially with respect to domain-specific aspects of EC education, given German students' poor results in the 2000 PISA wave (Ulferts & Anders, 2016). As a consequence, the 16 German federal states developed state-wide curricula for EC education that defined domain-specific cognitive objectives for the development of children (Diskowski, 2009) and revised their teacher education curricula accordingly to address domain-specific education (KMK, 2011, 2017; Robert Bosch Stiftung, 2008).

Since then, the federal states' teacher education curricula have taken a competence-oriented view on the outcomes of teacher education. In other words, the goal is to equip EC teachers with the domain-specific knowledge and skills they need to succeed in professional situations (Hartig & Klieme, 2006). Therefore, from a domain-specific point of view, pre-service EC teachers should develop professional knowledge, such as MPCK, as well as (domain-specific) situation-specific skills during teacher education (KMK, 2017; Robert Bosch Stiftung, 2008). However, there are indications that domain-specific aspects have yet to become a focus in teacher education (Gasteiger, Brunner et al., 2020; Kleeberger & Stadler, 2011). A closer look at the EC teacher education curricula for mathematics indicates that opportunities to learn mathematics are included on the level of school mathematics as well as in the form of EC mathematics. Descriptions within the curricula address MPCK as well as situation-specific skills. However, it is not clear how many opportunities to learn pre-service teachers are provided with respect to acquiring such knowledge and situation-specific skills. Furthermore, the curricula usually address early mathematics education together with other domains, such as early science education. This might lead to different opportunities to learn among pre-service EC teachers taught under the same curriculum.

Research question

Little is known about differences in situation-specific skills across students in different years of teacher education. Based on a random sample of pre-service EC teachers from the vocational track in Germany, we aim at providing first empirical indications regarding differences in situation-specific skills as well as MPCK among pre-service EC teachers in different years of teacher education. With regard to MPCK, the intention is to replicate existing findings (Blömeke et al., 2017; Torbeyns et al., 2019). We assume that German pre-service EC teachers' MPCK differs depending on which year of teacher education they are in. With respect to skills, the intention is to provide new findings. We assume that the skills in perceiving math-related situations and in planning math-related actions are likewise related to the year of teacher education and that pre-service EC teachers in a more advanced year of teacher education have developed greater skills in perceiving math-related situations and planning math-related actions than students in lower years.

Design and instruments

We investigated these research questions with a sample of $n = 334$ pre-service EC teachers from 16 vocational school classes in Lower Saxony and Berlin. The classes were at different stages of teacher education (first year = 41.5%, second year = 33.0% and third (last) year = 25.6%). Eighty-three per cent of the participants were female, which is in line with the population of pre-service EC teachers in Germany. The average age was $M = 23$ years (SD = 4 years).

To assess situation-specific skills, we used a video-based assessment documented by Dunekacke and colleagues (2015). The test consists of three short video clips. All video clips represent everyday life situations from a German EC institution recorded by the first author of the article. The selection of videos for the final assessment was based on theories of early mathematical learning (Clements & Sarama, 2007; Schuler, 2013) as well as expert ratings. The video clips address a range of mathematical content areas (Clements & Sarama, 2007) and include both teacher-guided (Video 1) and child-centred (Videos 2 and 3) activities. To assess situation-specific skills, participants answered a total of 24 mostly open-ended items. Of these, 12 address the skill in perceiving math-related situations and the other 12 assess the skill in planning math-related actions.

The skill in perceiving math-related situations was assessed for each video clip with the items "Does this situation contain mathematical themes?" (yes/no; one item) and "Please describe three aspects of this situation relevant from a mathematics education perspective and provide examples of evidence for each" (three items). The skill in planning math-related actions was assessed for each video clip with the items "Please provide two options on how to react appropriately in this situation from a mathematics education perspective" (two items) and "Please describe as precisely as possible two different possible ways to work on the theme in the coming days that would be appropriate from a mathematics education perspective" (two items). All open-ended items were coded dichotomously (right/wrong). The coding was based on a coding manual, with codes derived from an expert panel, the literature, and a small pre-test and further refined during the field study. Ten per cent of the data were coded twice by the first author and trained project staff. All items achieved a good inter-rater reliability (Yules $\Upsilon \geq$ .80). Cronbach's α reliability for the perception scale was .53 and thus rather low. The reliability for the planning action scale was .71 and thus satisfactory. Several studies have provided first evidence for the scale's validity with respect to inferences about pre-service EC teachers' skills (Dunekacke et al., 2015, 2016).

MPCK was assessed with a standardized test (Blömeke et al., 2017). The test represents four theoretically established aspects of early mathematical learning

Figure 7.1 Videos to assess situation-specific skills (Dunekacke et al., 2015).

You are playing a dice game with three children. Please explain, in short, why their mathematical learning in the following field is fostered: Numbers and operations (e.g. calculating):

_____

_____

Figure 7.2 Example item from the test to measure mathematics pedagogical content knowledge.

(organization of mathematical learning in planned and open situations; knowledge about the development of mathematical literacy, diagnosing mathematical literacy; fostering mathematical literacy among children with special needs). Several studies have provided evidence for the validity of the test (Blömeke, Jenßen et al., 2015). Figure 7.1 presents an example item from the test (translated into English).

For the present study, we used a short form of the MPCK test, which consisted of 12 items (Dunekacke et al., 2016) representing all aspects of early mathematical learning. The test consists of multiple-choice and open-ended items. An example is given in Figure 7.1. All items were coded dichotomously (right/ wrong). For the open-ended items, standardized coding manuals were used, and 10% of the data were double-coded ($\kappa \geq .88$). The short form of the test reached a satisfactory level of reliability, with $\alpha = .65$. All tests were administered by trained project staff within regular instructional time during teacher education. To answer our research question, we calculated an analysis of variance (ANOVA) to determine whether students in several years of teacher education differed in their MPCK, their skill in perceiving math-related situations or their skill in planning math-related actions.

Results

Table 7.1 presents the descriptive results for MPCK, perception of math-related situations (PERC) and planning math-related actions (ACT) for the overall sample as well as three years of teacher education separately. The range and standard

Table 7.1 Descriptive results

	Min	Max	M (SD)			
	All	All	All	First year	Second year	Third year
MPCK	0	12	7.3 (2.5)	6.8 (2.75)	7.2 (2.3)	8.2 (2.2)
PERC	1	11	6.4 (2.0)	5.6 (2.0)	6.5 (1.9)	7.4 (1.6)
ACT	0	11	4.3 (2.7)	3.6 (2.7)	4.2 (2.6)	5.6 (2.2)

ACT = skill in planning math-related actions, MPCK = Mathematics pedagogical content knowledge, PERC = skill in perceiving math-related situations.

derivations indicate that pre-service EC teachers in each year of teacher education vary widely in their knowledge and skills. Moreover, the data indicate that pre-service EC teachers may develop both knowledge and situation-specific skills over the course of teacher education.

In detail, we found significant differences between nearly all groups. Welch's test indicated small significant differences between the years of EC teacher education for MPCK ($F(2, 158.972) = 8.306$, $p < .001$, $\eta^2 = .06$). Because homogeneity of variance was violated, Games-Howell post-hoc analyses were chosen to test the differences between groups. The analysis showed a significant difference between first- and third-year students (-1.43, $p < .001$) as well as between second- and third-year students (-1.05, $p < .05$), but not between first- and second-year students (-0.38, $p = .57$).

For PERC, we found significant differences of medium size between the years of teacher education ($F(2, 258) = 20.115$, $p < .001$, $\eta^2 = .13$). Since homogeneity of variance could be confirmed, Tukey post-hoc analyses were applied and revealed differences between first- and second-year (-0.89, $p < .05$) as well as first- and third-year students (-1.76, $p < .001$) and between second- and third-year students (-0.87, $p < 05$).

For planning math-related actions, Welch's test again indicated significant differences of medium size between the three groups ($F(2, 155.546) = 15.539$, $p < .001$, $\eta^2 = .10$). Just like for MPCK, homogeneity of variance was violated; Games-Howell post-hoc analyses revealed significant differences between first- and third-year students (-1.98, $p < .001$) as well as second- and third-year students (-1.40, $p < .05$), but not between first- and second-year students (-0.58, $p = .27$), just as was the case for MPCK.

Discussion

Pre-service EC teachers' professional knowledge and especially their situation-specific skills have become an increasingly popular field of research in recent years. However, little is known about the acquisition of knowledge and skills within teacher education (Horm et al., 2013; La Paro et al., 2017). In this chapter, we investigated how pre-service EC teachers in several years of teacher education differ with regard to their MPCK, their skill in perceiving math-related situations in EC education and the skill in planning math-related actions within these situations.

Our results indicate that MPCK varies considerably, especially in the first two years of teacher training. However, it increases by the final year of teacher training. This is indicated by the ANOVA showing that students in the first and second year of teacher education did not differ in their MPCK, but that students in the last year of teacher training have achieved significantly more MPCK than students in the other two years. This replicates results by Blömeke and colleagues (2017), who showed that pre-service EC teachers at the beginning and the end of their teacher education program differed in their MPCK. However, the results are not in line with those of Torbeyns and colleagues (2019), who also found differences between pre-service EC teachers in their first year and those in their second or third year. This study took place in Belgium, where pre-service EC teachers may have more opportunities to learn.

Our results might indicate that opportunities to learn MPCK are placed more towards the end of EC teacher education in Germany. However, this interpretation needs to remain speculative because the German curricula give little or no indication of where opportunities to learn about domain-specific EC education, especially early mathematics education, are located within teacher education. If such learning opportunities indeed occur more at the end of teacher education, this might suggest that other aspects, such as general pedagogical knowledge, are addressed at the beginning of teacher education. Given the high correlation between general pedagogical knowledge and MPCK (Blömeke, Jenßen et al., 2015), an open question concerns whether it would be possible to teach the two in a more integrated way so that pre-service EC teachers can acquire greater and possibly also better connected MPCK. Research on teacher education has provided first evidence for such a result (Tröbst et al., 2018). However, further research – including research with experimental designs – will be necessary to clarify whether this is also a possibility for EC teacher education.

Regarding the skill in perceiving math-related situations, our results indicate that this pre-service EC teachers' skill differs across all three years of teacher education. This might indicate that pre-service EC teachers have opportunities to acquire this skill in all years of teacher education, in contrast to MPCK. Based on our data, we could not specify whether the differences in skill in perceiving math-related situations depend more on theory-based instruction or practical experiences within teacher education. As previous studies have demonstrated a close connection between the skill in perceiving math-related situations and MPCK (Dunekacke et al., 2016), it is an open question what the development of skill in perceiving math-related situations specifically refers to, especially at the beginning of training. Theoretical models of professional competence indicate that in addition to MPCK, other aspects like general pedagogical knowledge or beliefs might affect the acquisition of the skill in perceiving math-related situations (Blömeke, Gustafsson et al., 2015; Depaepe et al., 2020).

The pre-service EC teachers in our study differed in their skill in planning math-related actions depending on which year of teacher education they were in. However, the ANOVA results indicate that, like for MPCK, pre-service EC teachers' skill in planning math-related actions differs significantly at the end of teacher education but not at the beginning. Just as for MPCK, this might indicate that pre-service EC teachers have more opportunities to learn skills in planning math-related actions at the end of teacher education. Based on our data, we can again only speculate about what kinds of learning opportunities affect the acquisition of skill in planning math-related actions. Also in this case, an open question concerns whether opportunities to learn during instructional time in EC teacher education are responsible for the differences in skill in planning math-related actions, or if the skill in planning math-related actions is fostered by practical experiences within teacher education.

The range and standard derivations in Table 7.1 indicate that pre-service EC teachers in each year of teacher education vary widely in their knowledge and skills, with some able to successfully complete only a small number of items and some mastering nearly all tasks. Among students just beginning teacher education, this begs the question of where such knowledge was acquired. This is also an important issue for teacher educators from a practical perspective. They need to diagnose pre-service EC teachers' knowledge and skills and prepare math-related learning opportunities for heterogeneous groups of pre-service EC teachers. Furthermore, the results for students at the end of teacher education indicate that there might be a group of pre-service EC teachers who did not acquire the necessary knowledge and skills to master the items and hypothetically thus also the requirements of real-life teaching situations. Nevertheless, the results regarding all three dimensions of pre-service EC teachers' domain-specific competence indicate that an increase in the mean is accompanied by a decrease in the standard derivation. Cautiously interpreted, this might be a first indication that teacher education might be able to successfully reduce individual differences among pre-service EC teachers.

However, two limitations of the empirical study should be kept in mind. First, our data are of a cross-sectional nature. Therefore, it only allows for the investigation of differences over the three years of EC teacher education with convenience samples. Causal interpretations regarding the development of professional knowledge are not possible. Consequently, the results are of descriptive nature only. Second, a sample of only 16 classes from two German federal states was included in the investigation. Given that teacher education curricula differ across federal states, this sample is far from representative for Germany as a whole. We need in addition to assume that it is positively biased given the voluntary participation in our study. The results are therefore limited to first indications which may stimulate further research questions. Both research (Whyte, Stein, Kim, Jou, & Coburn, 2018) and discussions with teacher educators in the field indicate that the extent of math-related opportunities to learn depends not only on the federal state's curriculum but also on its implementation within each individual teacher education institution. This variation may be further increased by the fact that the corresponding curricula often combine opportunities to learn in different domains.

Future research should examine in more detail the question of opportunities to learn MPCK and especially skills in perceiving math-related situations and planning math-related actions within EC teacher education. This should include an examination of the amount of instructional time, or the frequency of opportunities to learn. It would also be necessary to focus on the quality of opportunities to learn. Particularly with regard to situation-specific skills and EC teachers' performance, it would be important to learn more about the relationship between opportunities to learn in teacher education and practical experience. Another challenge regarding the quality of opportunities to learn might be the fact that our results indicate a potentially large heterogeneity among pre-service EC teachers, especially at the beginning of teacher education. If this result were to be replicated in further research, an important task in the coming years will be to develop math-related teaching approaches to address these heterogeneous competence levels. An ongoing question will be how teacher education can support pre-service EC teachers' integration of different competence dimensions. Finally, a problem with the federal states' curricula is that a broad range of domains is to be addressed in a limited amount of instructional time. It would be worthwhile to explore whether there are similarities between different domains, such as mathematics and science, which could be used to make teacher education more effective across different domains of EC education.

References

Anderson, L. W., & Krathwohl, D. R. (Eds.) (2001). *A taxonomy for learning, teaching, and assessing: A revision of Bloom's taxonomy of educational objectives* (Abridged ed., [Nachdr.]). New York, NY: Longman.

Autorengruppe Fachkräftebarometer (2019). *Fachkräftebarometer Frühe Bildung 2019.* München: Deutsches Jugendinstitut; Weiterbildungsinitiative Frühpädagogische Fachkräfte. WiFF.

Ball, D. L., & Bass, H. (2009). With an eye on the mathematical horizon: Knowing mathematics for teaching to learners' mathematical futures. In M. Neubrand (Ed.), *Beiträge zum Mathematikunterricht 2009* (pp. 11–22). Münster: WTM.

Blömeke, S., Gustafsson, J.-E., & Shavelson, R. J. (2015). Beyond dichotomies: Competence viewed as a continuum. *Zeitschrift Für Psychologie, 223*(1), 3–13. https://doi.org/10.1027/2151-2604/a000194

Blömeke, S., Jenßen, L., Dunekacke, S., Suhl, U., Grassmann, M., & Wedekind, H. (2015). Leistungstests zur Messung der professionellen Kompetenz frühpädagogischer Fachkräfte. *Zeitschrift Für Pädagogische Psychologie, 29*(3–4), 177–191. https://doi.org/10.1024/1010-0652/a000159

Blömeke, S., Jenßen, L., Grassmann, M., Dunekacke, S., & Wedekind, H. (2017). Process mediates structure: The relation between preschool teacher education and preschool teachers' knowledge. *Journal of Educational Psychology, 109*(3), 338–354. https://doi.org/10.1037/edu0000147

Bruns, J., & Gasteiger, H. (2019). VIMAS_NUM: Measuring situational perception in mathematics of early childhood teachers. In M. Graven, H. Venkat, A. Essien, & P. Vale (Eds.), *Proceedings of the 43rd conference of the International Group for the Psychology of Mathematics Education* (pp. 129–136). Pretoria, South Africa: PME.

Bruns, J., Strahl, C., & Gasteiger, H. (2020). Situative Beobachtung und Wahrnehmung angehender frühpädagogischer Fachpersonen im Bereich Mathematik – Entwicklung und Validierung eines Testinstruments. *Unterrichtswissenschaft, 31*(2), 255. https://doi.org/10.1007/s42010-020-00091-7

Clements, D. H., & Sarama, J. (2007). Early childhood mathematics learning. In F. K. Lester (Ed.), *Second handbook of research on mathematics teaching and learning* (pp. 461–555). Charlotte, NC: Information Age Publ.

Depaepe, F., Verschaffel, L., & Kelchtermans, G. (2013). Pedagogical content knowledge: A systematic review of the way in which the concept has pervaded mathematics educational research. *Teaching and Teacher Education, 34*, 12–25. https://doi.org/10.1016/j.tate.2013.03.001

Depaepe, F., Verschaffel, L., & Star, J. (2020). Expertise in developing students' expertise in mathematics: Bridging teachers' professional knowledge and instructional quality. *ZDM, 52*(2), 179–192. https://doi.org/10.1007/s11858-020-01148-8

Diskowski, D. (2009). Bildungspläne für Kindertagesstätten – Ein neues und noch unbegriffenes Steuerungsinstrument. In H.-G. Roßbach & H.-P. Blossfeld (Eds.), *Zeitschrift für Erziehungswissenschaft Sonderheft: Vol. 11. Frühpädagogische Förderung in Institutionen* (pp. 47–61). Wiesbaden: VS Verlag für Sozialwissenschaften/GWV Fachverlage GmbH Wiesbaden. https://doi.org/10.1007/978-3-531-91452-7_4

Dunekacke, S., Jenßen, L., & Blömeke, S. (2015). Effects of mathematics content knowledge on pre-school teachers' performance: A video-based assessment of perception and planning abilities in informal learning situations. *International Journal of Science and Mathematics Education, 13*(2), 267–286. https://doi.org/10.1007/s10763-014-9596-z

Dunekacke, S., Jenßen, L., Eilerts, K., & Blömeke, S. (2016). Epistemological beliefs of prospective preschool teachers and their relation to knowledge, perception, and planning abilities in the field of mathematics: A process model. *ZDM, 48*(1–2), 125–137. https://doi.org/10.1007/s11858-015-0711-6

Fröhlich-Gildhoff, K., Nentwig-Gesemann, I., & Pietsch, S. (2011). *Kompetenzorientierung in der Qualifizierung frühpädagogischer Fachkräfte.* Deutsches Jugendinstitut, München.

Gasteiger, H., & Benz, C. (2018). Enhancing and analyzing kindergarten teachers' professional knowledge for early mathematics education. *The Journal of Mathematical Behavior, 51*, 109–117. https://doi.org/10.1016/j.jmathb.2018.01.002

Gasteiger, H., Brunner, E., & Chen, C.-S. (2020). Basic conditions of early mathematics education—a comparison between Germany, Taiwan and Switzerland. *International Journal of Science and Mathematics Education, 16*(3). https://doi.org/10.1007/s10763-019-10044-x

Gasteiger, H., Bruns, J., Benz, C., Brunner, E., & Sprenger, P. (2020). Mathematical pedagogical content knowledge of early childhood teachers: A standardized situation-related measurement approach. *ZDM, 52*, 193–205. https://doi.org/10.1007/s11858-019-01103-2

Hartig, J., & Klieme, E. (2006). Kompetenz und Kompetenzdiagnostik. In K. Schweizer (Ed.), *Leistung und Leistungsdiagnostik: Mit 18 Tabellen* (pp. 127–143). Berlin, Heidelberg: Springer Medizin Verlag Heidelberg. https://doi.org/10.1007/3-540-33020-8_9

Horm, D. M., Hyson, M., & Winton, P. J. (2013). Research on early childhood teacher education: Evidence from three domains and recommendations for moving forward. *Journal of Early Childhood Teacher Education, 34*(1), 95–112. https://doi.org/10.1080/10901027.2013.758541

Kersting, N. B., Givvin, K. B., Sotelo, F. L., & Stigler, J. W. (2010). Teachers' analyses of classroom video predict student learning of mathematics: Further explorations of a novel measure of teacher knowledge. *Journal of Teacher Education, 61*(1–2), 172–181. https://doi.org/10.1177/0022487109347875

Kleeberger, F., & Stadler, K. (2011). *Zehn Fragen – zehn Antworten: Die Ausbildung von Erzieherinnen und Erziehern aus Sicht der Lehrkräfte; Ergebnisse einer bundesweiten Befragung von Lehrkräften an Fachschulen für Sozialpädagogik; eine Studie der Weiterbildungsinitiative Frühpädagogische Fachkräfte (WiFF). WiFF-Studien Ausbildung (Vol. 13).* München: DJI.

KMK (2011). *Kompetenzorientiertes Qualifikationsprofil für die Ausbildung von Erzieherinnen und Erziehern an Fachschulen/Fachakademien: Beschluss der Kultusministerkonferenz vom 01.12.2011.*

KMK (2017). *Kompetenzorientiertes Qualifikationsprofil für die Ausbildung von Erzieherinnen und Erziehern an Fachschulen und Fachakademien: Beschluss der Kultusministerkonferenz vom 01.12.2011 i.d.F. vom 24.11.2017.*

La Paro, K. M., van Schagen, A., King, E., & Lippard, C. (2017). A systems perspective on practicum experiences in early childhood teacher education: Focus on interprofessional relationships. *Early Childhood Education Journal, 18*(2), 31. https://doi.org/10.1007/s10643-017-0872-8

Lee, J. (2010). Exploring kindergarten teachers' pedagogical content knowledge of mathematics. *International Journal of Early Childhood, 42*(1), 27–41.

Linder, S. M., & Simpson, A. (2018). Towards an understanding of early childhood mathematics education: A systematic review of the literature focusing on practicing and prospective teachers. *Contemporary Issues in Early Childhood, 19*(3), 274–296. https://doi.org/10.1177/1463949117719553

Lindmeier, A., Hepberger, B., Heinze, A., & Moser Opitz, E. (2016). Modeling cognitive dispositions of educators for early mathematics education. *Proceedings of the 40th Conference of the International Group for the Psychology of Mathematics Education, 3,* 219–226.

Lindmeier, A., Seemann, S., Kuratli-Geeler, S., Wullschleger, A., Dunekacke, S., Leuchter, M., Vogt, F., Moser Opitz, E., & Heinze, A. (2020). Modelling early childhood teachers' mathematics-specific professional competence and its differential growth through professional development – An aspect of structural validity. *Research in Mathematics Education, 22*(2), 168–187. https://doi.org/10.1080/14794802.2019.1710558

McCray, J. S., & Chen, J.-Q. (2012). Pedagogical content knowledge for preschool mathematics: Construct validity of a new teacher interview. *Journal of Research in Childhood Education, 26*(3), 291–307. https://doi.org/10.1080/02568543.2012.685123

Meschede, N., Fiebranz, A., Möller, K., & Steffensky, M. (2017). Teachers' professional vision, pedagogical content knowledge and beliefs: On its relation and differences between pre-service and in-service teachers. *Teaching and Teacher Education, 66*, 158–170. https://doi.org/10.1016/j.tate.2017.04.010

Oberhuemer, P., Schreyer, I., & Neuman, M. J. (2010). *Professionals in early childhood education and care systems: European profiles and perspectives.* Opladen: Budrich.

Oppermann, E., Anders, Y., & Hachfeld, A. (2016). The influence of preschool teachers' content knowledge and mathematical ability beliefs on their sensitivity to mathematics in children's play. *Teaching and Teacher Education*, *58*, 174–184. https://doi.org/10.1016/j.tate.2016.05.004

Paprzycki, P., Tuttle, N., Czerniak, C. M., Molitor, S., Kadervaek, J., & Mendenhall, R. (2017). The impact of a framework -aligned science professional development program on literacy and mathematics achievement of K-3 students. *Journal of Research in Science Teaching*, *97*(1), 830. https://doi.org/10.1002/tea.21400

Robert Bosch Stiftung (2008). *Frühpädagogik Studieren – ein Orientierungsrahmen für Hochschulen*.

Santagata, R., & Yeh, C. (2016). The role of perception, interpretation, and decision making in the development of beginning teachers' competence. *ZDM*, *48*(1–2), 153–165. https://doi.org/10.1007/s11858-015-0737-9

Schuler, S. (2013). *Mathematische Bildung im Kindergarten in formal offenen Situationen: Eine Untersuchung am Beispiel von Spielen zum Erwerb des Zahlbegriffs* (1. Aufl.). Münster: Waxmann Verlag GmbH.

Shulman, L. S. (1986). Those who understand: Knowledge growth in teaching. *Educational Researcher*, *15*(2), 4–14.

Stahnke, R., Schueler, S., & Roesken-Winter, B. (2016). Teachers' perception, interpretation, and decision-making: A systematic review of empirical mathematics education research. *ZDM*, *48*(1–2), 1–27. https://doi.org/10.1007/s11858-016-0775-y

Star, J. R., & Strickland, S. K. (2008). Learning to observe: Using video to improve preservice mathematics teachers' ability to notice. *Journal of Mathematics Teacher Education*, *11*(2), 107–125. https://doi.org/10.1007/s10857-007-9063-7

Torbeyns, J., Verbruggen, S., & Depaepe, F. (2019). Pedagogical content knowledge in preservice preschool teachers and its association with opportunities to learn during teacher training. *ZDM*, *29*(3), 305. https://doi.org/10.1007/s11858-019-01088-y

Tröbst, S., Kleickmann, T., Heinze, A., Bernholt, A., Rink, R., & Kunter, M. (2018). Teacher knowledge experiment: Testing mechanisms underlying the formation of preservice elementary school teachers' pedagogical content knowledge concerning fractions and fractional arithmetic. *Journal of Educational Psychology*, *110*(8), 1049–1065. https://doi.org/10.1037/edu0000260

Ulferts, H., & Anders, Y. (2016). Effects of ECEC on academic outcomes in literacy and mathematics: Meta-analysis of European longitudinal studies. *Manuscript submitted for publication*. http://ecec-care.org/fileadmin/careproject/Publications/reports/CARE_WP4_D4_2_Metaanalysis_public.pdf (accessed 25.01.2021).

van Es, E. A., & Sherin, M. G. (2008). Mathematics teachers' "learning to notice" in the context of a video club. *Teaching and Teacher Education*, *24*(2), 244–276. https://doi.org/10.1016/j.tate.2006.11.005

Whyte, K. L., Stein, M. A., Kim, D., Jou, N., & Coburn, C. E. (2018). Mathematics in early childhood: Teacher educators' accounts of their work. *Journal of Early Childhood Teacher Education*, *39*(3), 213–231. https://doi.org/10.1080/10901027.2017.1388306

Wittmann, G., Schuler, S., & Levin, A. (2015). To what extent can kindergarten teachers and primary school teachers initiate and foster learning mathematics in typical situations? In *CERME 9 – Ninth Congress of the European Society for Research in Mathematics Education*. https://hal.archives-ouvertes.fr/hal-01289887/document (accessed 25.01.2021).

Chapter 8

Supporting pre-service early childhood educators to identify mathematical activities in the actions of preverbal young children

Audrey Cooke, Jenny Jay

Introduction

In initial teacher education (ITE) programmes for pre-service early childhood educators in Australia, one focus in the area of mathematics is to develop educator understanding of how early mathematics is experienced by very young children (birth–2 years). Therefore, it is critical to build educator confidence and capacity to recognise and create opportunities for young children to engage with mathematical activities in their learning environment to maximise foundational mathematical learning. This involves pre-service early childhood educators potentially rethinking what they believe mathematics is, who may engage with mathematics, and what engagement might look like.

The pre-service educators' attitudes are vital in this process – instead of viewing mathematics a process or activity that is restricted to specific circumstances or environments, viewing it as something people engage with and use everyday (Freudenthal, 1968, p. 4). Mathematising is considered by van Oers (2014, p. 114) to be evident in what young children do when playing. Mathematics is therefore not limited to the formalised processes developed and conducted in school and university classrooms, rather something that young children use when engaging with the world (van Oers, 2014, p. 113) – and young children are choosing to use these processes, even without knowledge of it having the name of "mathematics".

van Oers views mathematising as being embedded within the child's cultural context (2014, p. 114). Björklund (2008, p. 85) also makes the observation that mathematics is within the child's culture and considers the child's innate ability as impacted by the opportunities presented in their environment and cultural settings. Bishop (1988, p. 182) considers mathematics as a product of culture and a "cultural phenomenon" (Bishop, 1991, p. 3) and was able to identify his six mathematical activities from the "cultural knowledge" – counting, locating, measuring, designing, playing, and explaining (1988, pp. 182–183).

The following sections of this chapter explore young children's engagement with mathematics. How Bishop's (1988, pp. 182–183) mathematical activities can be a framework to provide a structure that can be used by early childhood educators to identify young children's mathematical thinking is considered.

DOI: 10.4324/9781003172529-8

Particular attention is given here to the group identified as "preverbal children" in order to clarify where mathematical activity can be identified without an oral explanation. The role of the early childhood educator is then considered and this is extrapolated to the development of pre-service early childhood educators, with specific focus on developing the situation-specific skill of identifying preverbal children's mathematical thinking. This involves a process of viewing very young children's behaviour in play several times and breaking it down into its individual actions. This deconstruction of behaviours into actions will give the pre-service educators an opportunity to see the complexity of mathematical thinking and understanding involved in a quite short sequence of behaviours.

Theoretical framework and understandings

This work looks at the child's engagement with their environment. Two theoretical approaches have been used to support the work in this chapter. Although they are from different perspectives, both consider the child as the main focus – which is what makes them supportive in this work. The first theoretical approach from Bronfenbrenner (1977) considers how that which surrounds the child – the environment within which they interact – may impact on the child. This environment influences the child and can both expand and constrain what activities the child may engage with. The second theoretical approach which supports the work of this chapter is from Franzén (2015), which focuses upon the child's engagement with their environment. However, as the child is very young and preverbal, it is their actions that are of interest.

Bronfenbrenner's (1977) theoretical framework of the ecology of human development provides a structure to consider how individuals interact with the environment where they live, grow and learn their culture. Through an examination of the close environment in which individuals live and their "larger social contexts", it is possible to focus on one aspect of a child's growing mathematical understanding. The contexts for this study are the child themselves, what they bring to a learning context and the settings where young children spend their times, such as early learning centres. While Bronfenbrenner (1977) describes the ecological theory as set of environments, each impacting upon the other, this chapter will look at the child in a community setting outside the home environment. These settings in which the observations (videos) take place show the impact of the child's immediate environment contained within the care setting.

Franzén, in 2015, drew on the work of Barad (as cited in Franzén, 2015, p. 46) in which language and its power is de-emphasised when observing very young children. Removing the power of language strengthens that which is observed, thus allowing for the children's actions, as they engage with their environment, to be examined and discussed. It is the close consideration of the children's actions that can give insight into the cognitions that the children may be using as they engage with their environment. The children's actions can be linked to the mathematical activities suggested by Bishop (1988).

Young children engaging with mathematics

Freudenthal (1968) saw mathematics as an activity, where an individual engages in "the process of mathematising reality" (p. 7), making connections between what happens in their world and the more formalised mathematics that may occur as part of their education and vice versa. Mathematising was viewed by van Oers (2014) as a way to learn about the world and, as such, he considered it "embedded in young children's play" (p. 114). Young children engage with mathematical thinking in their everyday lives and in the planned mathematically focused activities of their learning environments (Björklund, 2008). Preverbal young children, although unable to verbally explain their thinking, can show their mathematical understandings via their actions (Björklund, 2008).

Contrary to the view that mathematics is only used by a specialised few, mathematics is used and needed by almost everyone (Freudenthal 1968, p. 5) in their daily lives and their everyday experiences. Freudenthal (1968, p. 4) describes mathematics as useful, an endeavour that we all can use, but he argued that learners need to have the opportunities to experience mathematics as being useful – not just in mathematics but beyond, in their other schooling and out-of-school contexts. Freudenthal (1968) proposed that it is important for educators to understand the process of "mathematising", where mathematics is understood as an activity and a process that is used to understand and engage with the world (p. 7).

The idea of mathematising was discussed further by van Oers (2014, p. 113). He built on Freudenthal's (1968, p. 7) focus by incorporating the problem-solving actions of the individual undertaken to learn about the world they live in. van Oers (2014) adds that this made mathematics a form of play (p. 114), something children do as they quantify elements of their world while engaging with it. Whilst engaging in the world and experiencing the world, young children build their understandings of the world. These experiences continually build upon each other, becoming the basis of further understandings (Björklund, 2008, p. 85). As young children learn, they transfer their learnings of the world to other contexts they experience – sometimes successfully and sometimes not – a process that van Oers (2014, p. 114) considers invaluable in mathematising.

Both van Oers (2014, p. 114) and Björklund (2008, 2012, p. 216) emphasise the importance of the young child's environment in providing opportunities for them to engage with the world. The environment is what the child explores, and what they participate within whilst mathematising. The young child's cultural world impacts both the child and their environment including communication opportunities, objects they are provided with, and the values embraced (van Oers, 2014, p. 120). Likewise, Björklund (2008, p. 85) describes how culture can drive the "tools that make everyday life a bit easier" – these tools often involve mathematics, and can become the impetus for the young child's mathematical competencies. She proposes that these competencies focus the mathematising that the young child undertakes when engaging with the world.

Play provides opportunities for children to engage with mathematics on many levels. Björklund (2012, p. 217) explains how play is important for young children as it provides context that affords both a common interest for the children and is meaningful to them. Children use play as a medium for exploration and discovery, practice of new and emerging skills, and in being child-initiated, it is always meaningful to the child (Björklund, 2012, p. 217) making it a productive and positive context for young children to develop and deepen mathematical concepts.

In his seminal work Bishop (1991, p. 3) states "mathematics is a cultural phenomenon". He considers mathematics as having been "conceived of as a cultural product, which has developed as a result of various activities" (Bishop, 1988, p. 182). Bishop (1991, p. 22) describes communication as occurring in all cultures, with what is communicated being important to those cultures. He made the parallel proposal that all cultures also engage in mathematical activities, suggesting that just as language is developed within cultures, so is mathematics (1991, p. 22). In proposing this, Bishop (1991, p. 23) emphasised that culture embraces the individuals, the social interactions, and the environments, and these all motivate and stimulate opportunities for mathematical thinking, ideas, and engagement.

Although Bishop's (1988, 1991) mathematical activities help identify mathematical understandings in verbal children, difficulties arise when attempting to use them to identify mathematical understandings and thinking of preverbal very young children (birth–2 years). This is not an indication that preverbal young children do not engage with mathematical thinking, rather that their engagement with mathematics is not evident through language but through their body's actions (Meaney, 2016). In her investigation of very young children engaged in play, Franzén (2015) identifies the mathematical thinking that could be evident from the actions of a preverbal young child during her interaction with a toy climb-in car. The observed actions show that preverbal young children engage in mathematical thinking when they are investigating their world.

Bishop (1988, 1991) describes six mathematical activities – counting, locating, measuring, explaining, designing, and playing – based on language. These mathematical activities can be effectively used to help educators identify the mathematical understandings of verbal children. Bishop's (1988, pp. 182–183) six mathematical activities are counting, where discrete phenomena are itemised, compared, or ordered; locating, where the environment is explored, navigated, and conceptualised; measuring, where non-countable phenomena are quantified, compared, or ordered; designing, where the components of the environment are identified in terms of shape, template, or design; playing, where rules or processes are devised or formalised for activities; and explaining, where phenomena are described or analysed. He stated that there could be overlaps between the mathematical activities and that several may occur at once (Bishop, 1991, p. 108), reflecting the connected nature of mathematics in context. However, the conscious and sustained engagement with these mathematical activities would create the "cultural knowledge" underpinning mathematics (Bishop, 1988, p. 183).

Björklund (2008, p. 92) proposes that young children need to have multiple experiences in diverse environments to enable the young child to value their mathematical thinking and the many ways mathematics can be used. They need to be able to share what they know, but in a way that makes sense for them, not enforced or limited by adults but rather with the adults supporting them (Björklund, 2008, p. 93). It is the opportunities for young children to choose their activities, to engage with play, and to interact with the others in the environment that enables the development of mathematical thinking (Björklund, 2012, p. 215).

Young children's language skills are not a prerequisite for engaging with mathematical activities. In their research, Zippert, Eason, Marshall, and Ramani (2019, p. 10) found that children's language skills did not impact their exploration of mathematical-focused toys, but their mathematical understandings and knowledge did. When observing pairs of young children, they noticed the children explored the mathematical toys both verbally and nonverbally, with twice as much nonverbal explorations occurring during the activity. However, the researchers did caution that their results regarding verbal exploration may be due to the measures used to record their results.

Although crafted from language, Bishop's (1991, p. 22) mathematical activities are valuable for use with young children. Johansson, Lange, Meaney, Riesbeck, and Wernberg (2016, p. 28) outline how Bishop's mathematical activities support the Swedish pre-school curriculum. Meaney (2016, p. 11) used Bishop's mathematical activity of locating when investigating the mathematical thinking of young children who demonstrated understanding through their actions rather than their language. Adults who work with young children need to recognise that it is not just language that is used to learn and to communicate, but it is also the child's full-body engagement (Franzén, 2015, p. 47). This places emphasis on the educators' engagement with mathematics but, much as Björklund (2012, p. 215) contemplates, educators are sometimes unaware of the impact of their role, especially when their conceptions of mathematics may not be commensurate with creating opportunities for young children to engage mathematically with their world.

Bishop's mathematical activities and young children

Bishop's (1988, pp. 182–183) mathematical activities have been used by researchers to investigate the mathematical thinking that young children may engage with. In their research, Johansson et al. (2016, pp. 32–33) found that children at a Swedish pre-school engaged with all of Bishop's (1988, pp. 182–183) mathematical activities, often engaging in more than one at a time. However, they found that the educator could impact on mathematical learning through the provision of a rich environment and through asking thoughtful questions. This potential impact places emphasis on both the early childhood educator and the learning opportunities provided for pre-service early childhood educators.

Bishop (1991, p. 22) states, "it is now well established that all human groups communicate, and also that all cultures develop language". However, language is not the only way that people communicate nor share mathematical under-standings – it may be that language is prominent, but the dominance is not deserved (Barad, 2007, p. 132), that "language has been granted too much power" (Barad, 2007, p. 131). Björklund (2008, p. 85) argues that children who are preverbal can demonstrate their understandings via actions. Likewise, Meaney (2016, p. 5) criticised theories which highlight language as the convey-ance of mathematics. In her observation of a very young child, Franzén (2015, p. 52) found that the child physically engaged with her learning and explora-tion of the world by using her body to experience the mathematics during her explorations. In observing the child and connecting her actions to mathematical thinking (without obtaining verbal expressions or confirmations), Franzén con-nected the child's physical actions to mathematics, using the child's positioning and movement of her body as an expression of mathematical thinking (p. 52). Using Bishop's (1988, pp. 182–183) mathematical activities as part of the ITE programme for pre-service early childhood educators provides a structure with specific examples they can use when observing preverbal children's actions. This is beneficial as it helps clarify what mathematical thinking the preverbal children might be using when engaging in the observed actions.

The role of the early childhood educator

Early childhood educators impact on the opportunities young children have to engage mathematically. Stites and Brown (2019, p. 14) considered the role of the early childhood educator in developing young children's mathemat-ical thinking as one of facilitation and modelling, which should be under-taken when opportunities present themselves in the learning environment. However, to do this, early childhood educators need the situation-specific skill of being able to identify mathematics, both evident in the situation and in the mathematics that may be developed from the situation. This enables the early childhood educator to create appropriate opportunities for young children to mathematise through the use of mathematics in their everyday experiences (Björklund, 2018, p. 42).

Although mathematics in and of itself is not visible when engaged with, Björklund and Pramling (2017, p. 68) explain that it can be seen through observing young children's explorations and interactions with their environment and the objects (including humans) in their environment. These observations underpin the capacity to identify mathematics and form the basis of decisions made by the early childhood educator, particularly those that involve "directing the learner's attention to experienced mathematical relationships and supporting further exploration of that relationship" (Björklund & Pramling, 2017, p. 68). These decisions then create the environment that will enable the young children to continue to engage with mathematics.

The capacity to plan and to make planning decisions are seen by Björklund (2012, p. 218) as part of the early childhood educators' knowledge and skills. She proposes that planning and decisions need to identify experiences young children can engage with that will develop and extend their mathematical understandings and skills (2018, p. 38). To do this, the experiences need to be suitable for the young children, capturing their attention and developing their interest, and enable mathematical learning to occur (2012, p. 218), by building on the previous experiences and the mathematical understandings and knowledge that have been developed from them (2012, p. 217).

In their decisions and planning, early childhood educators need to focus on the opportunities young children have for learning (Franzén 2015, p. 50). Franzén (2015, p. 50) proposes that the starting point for early childhood educators when planning is to identify the children's underlying understandings. To do this, the educator must be able to identify the mathematics in what the young children do – and this becomes critical when young children are pre-verbal, because oral language is not available as a medium to communicate the young child's mathematical thinking. Palmér, Henriksson, and Hussein (2016, p. 80) state that educators need to have pedagogical and content knowledge about how children think about, perceive, and learn mathematics. In addition, they argue that educators who do not know these important elements consider children's development of mathematics to be either without educator input and planning or requiring tightly enacted and structured adult-directed experiences. This points to the importance of knowing what educator knowledge, understandings, and skills are needed (Blömeke, Gustafsson, & Shavelson, 2015, p. 8). Cohrssen and Tayler (2016, p. 35) emphasise this even further, asserting that the capacity of the educator to identify the mathematics young children already know is essential.

The Australian context

In Australia, the Australian Government Department of Education and Training [AGDET] (2009, p. 5) publish the Early Years Learning Framework (EYLF), a document that "forms the foundation for ensuring that children in all early childhood education and care settings experience quality teaching and learning". The EYLF both guides and underpins decisions early childhood educators make regarding the curriculum (p. 8). Principles, practice, and learning outcomes are considered by the EYLF as central, impacting on "all the interactions, experiences, routines and events, planned and unplanned, that occur in an environment designed to foster children's learning and development" (AGDET, 2009, p. 9). The EYLF also values children's agency and recognises their active construction of their understandings and encourages educators to discard preconceptions about children's capabilities in favour of acknowledging the uniqueness of each child (AGDET, 2009, p. 9). The identification of young children's mathematical capabilities – both in terms of what they can do and what they are able to learn and develop – are provided within two of the five learning outcomes (AGDET, 2009, pp. 35, 38). Cooke (2018, p. 6) identifies specific key components within these two learning outcomes as either explicitly or potentially empowering the children's engagement with mathematics in a way that is both confident and creative.

Australia also has the Australian Institute for Teaching and School Leadership (AITSL) that produces the Australian Professional Standards for Teachers (APST). The APST provides seven standards organised into three domains of teaching that address educator knowledge, practice, and educator engagement (AITSL, 2011, p. 4). Developing children's mathematical understandings and numeracy is specifically addressed within Standard 2, "know the content and how to teach it" (AITSL, 2011, p. 12) under the focus area "literacy and numeracy strategies" (AITSL, 2011, p. 13), connecting the educator's capacity to identify and use effective teaching strategies that enable the children they work with to develop numeracy. The APST are used by AITSL to accredit ITE programmes, including those for early childhood educators, in Australia (AITSL, 2011, p. 6).

The development of pre-service early childhood educators

Early childhood educator ITE programmes in Australia are responsible for the development of pre-service early childhood educators who graduate with a specialisation specific to children from birth to 8 years of age. The ITE programme produces early childhood educators who can create and deliver what Cohrssen and Tayler (2016, p. 37) describe as "high-quality, play-based education and care programs that both meet the current interests and learning needs of individual children, but which also prepare children for the transition to formal school education". To get to this point, the early childhood educator ITE programmes must provide opportunities for the pre-service early childhood educator to develop and become competent in the skills and understandings that they will need. Although Blömeke et al. (2015, pp. 3, 7) consider competence to incorporate disposition, both cognitive and affective-motivational; situation-specific skills connected to perceptions, interpretations, and decision-making; and the performance that is evident in the observable behaviours, this chapter focuses on the skill of identifying mathematics in what preverbal children do.

Cohrssen and Tayler (2016, p. 35) state that identifying children's current mathematical understandings and then using those understandings to inform planning are key capabilities for early childhood educators, but they warned that pre-service early childhood educators often have a lack of confidence or self-efficacy that can impede their development (p. 26). One of the points they noted was the focus on how important it is to carefully observe and plan to support children's development of counting skills (p. 36), as well as the assistance that fine-grain observations provided to the pre-service early childhood educators (p. 37). Cohrssen and Tayler (2016, p. 25) reiterate that early childhood ITE programmes should ensure they prepare pre-service early childhood educators to be able to engage and support young children mathematically. This role, they stated, should also consider the pre-service early childhood educators' mathematical self-efficacy, mathematics anxiety, attitudes towards mathematics, and beliefs about mathematics.

The skill of identifying mathematics in young children's behaviour and actions may be hampered by the pre-service early childhood educators' willingness to engage with mathematics. Bates, Latham, and Kim (2013, p. 8) described how pre-service early childhood educators have different fears related to mathematics and to teaching mathematics, with mathematical content itself creating fear as well as being why they were afraid of mathematics. As a consequence, they proposed that early childhood educator ITE programmes need to engage pre-service early childhood educators in activities that increase their exposure to mathematics in a way that reduces their fear (p. 9).

Identifying the mathematics in what preverbal children do

To develop the skill of recognising and identifying the mathematical thinking in preverbal children's behaviours and actions, early childhood educator ITE programmes need to provide opportunities for pre-service early childhood educators to see the mathematics in what young children do (Meaney, 2016, p. 5), but in a way that the pre-service early childhood educators are comfortable with (Bates et al., 2013, p. 9). This section of this chapter considers how Bishop's (1988, pp. 182–183) mathematical activities can be used, but with a de-emphasis on language (Barad, 2007, p. 131; Björklund, 2008, p. 85). The affordances made possible by video – the capture of the real-time experience (Lynch & Stanley, 2018, p. 63) and the iterative nature of viewing the video (Björklund, 2008, p. 86; de Freitas, 2016, p. 554) – are discussed as tools that create optimal experiences for pre-service early childhood educators. These tools are then considered in an approach developed for pre-service early childhood educators in an ITE programme at an Australian university. The limitations are provided within the discussion of the approach.

Video affordances

In their work investigating the occupations of very young children in their home environments, Lynch and Stanley (2018, p. 63) found video was very effective in capturing both the social and physical environments, with the affordances surpassing those provided through still photographs and interviews. Importantly for work with preverbal children that is focusing on actions, they noted that video enabled "a unique way to study occupation beyond words, to record events as they unfold in their physical, temporal, and social contexts" (Lynch & Stanley, 2018, p. 58). They further describe how very young children interact with their environment, referring to it as "the unfolding developmental relationship between the infant and the physical environment" (p. 63). This reflects Barad's (2007, p. 74) focus on the intra-action – which, rather than a one-way interaction, creates changes in both the components of the intra-action (in this instance, the child and the physical environment) as well as between them. Video enables these to be captured.

Björklund (2008, p. 86) identifies the opportunity to watch what is occurring within the full environment and to do so in a comprehensive manner. Unlike watching in real time, video captures everything that happens. The use of 360-degree video with the placement of the camera close to the ceiling enables vision of the full room to be recorded. All perspectives captured by the video can be considered, which enables the observer to follow the actions and behaviours of all in the room (Björklund, 2008, p. 86). Being within the environment but unobtrusive, everyday life and contexts present themselves, which may not be possible with a researcher in the room with the participants (Björklund, 2008, p. 87). As the focus is on the skill of the pre-service early childhood educator identifying the mathematics in what preverbal young children do, the affordances presented through the use of 360-degree video enables the unobtrusive and repeated observations of very young children in their everyday environment (Björklund, 2008, p. 88).

The affordances generated through the use of video involving the creation of new knowledge and of new ways of knowing were identified by de Freitas (2016, p. 554). She focused on the opportunities to capture the interactions between the participants, highlighting the capacity for this to profoundly change research in educational settings. It enables closer examination of how the use of the body might communicate mathematical thinking that is not otherwise visible or audible (de Freitas, 2016, p. 564). Meaney (2016, p. 19) noted the affordance of seeing how very young children physically solve problems, particularly ones that are relevant to their context and environment. Utilising video within the ITE programme for pre-service early childhood educators provides affordances not otherwise available as many children are seen engaging with their everyday environment without the imposition of the pre-service early childhood educator within that environment. The 360-degree video also offers multiple perspectives that the viewer can take, which enables a greater recognition of the full environment that the young children are part of.

An iterative process

The iterative process enabled through the use of video provides the opportunities for multiple observers to view and re-view the video (Lynch & Stanley, 2018, p. 63), whether at the same time, together or separated, or temporally separated. These viewings and re-viewings can help the observers identify more from the actions, behaviours, and engagements that may not have otherwise been noticed (Lynch & Stanley, 2018, p. 63). Björklund (2008, p. 86) also acknowledges the opportunities to view video repeatedly, enabling observers to catch moments that may otherwise have been lost. She also highlights the prospect of a more thorough consideration of what is viewed, which could "allow for a more thorough analysis of content and meaning in a given situation and ensure that different perspectives are taken into account" (2008, p. 86). Most pertinent for the pre-service childhood educators in developing the skill of identifying the mathematics in what preverbal young children do, Björklund (2008, p. 86) proposes that the iterative process enabled through video "gives the researcher an opportunity to follow, analyse and interpret even the youngest child's learning process". In her research, videos were considered as an important tool in being able to focus on all participants within the environment, to see the actions of each but also the combined actions of all (Björklund, 2008, p. 87).

The use of the 360-degree video enables a continual re-viewing of the video by all participants within the ITE programme – both the pre-service early childhood educators and the educator who is working with them. The full environment can be viewed multiple times and specific sections focused upon. Observations can be noted, considered and reflected upon by the individual, and then discussed with peers and educators in both a synchronous and asynchronous format.

An approach developed for pre-service early childhood educators

The approach uses a 360-degree video platform. The 360-degree videos are captured in learning centre rooms where children are under the age of three years, with the majority preverbal. Care is taken to protect participants (Björklund, 2008, p. 87), such as providing access to participants (likely educators in the room who know the children) to identify sections of the video not to be shown to others. Knowing video is being recorded can impact on the actions of those who are aware, particularly adult participants, and this may result in less natural occurrences being captured (Lynch & Stanley, 2018, p. 58). Lynch and Stanley (2018, p. 58) note the possibility of power and influence to occur with video. These can relate to how decisions are made regarding what sections of video are viewed – similar to de Freitas (2016, p. 556), who highlights the consideration of the video not shown, which can create potential bias in what video is selected to view. To reduce the impact of these limitations, the selection of videos occurs after careful consideration of what is available. Short vignettes that have been selected by the educators provide a clearly visible and contained episode with a potential range of mathematical activities (Bishop, 1988, pp. 182–183) that can be identified.

The pre-service early childhood educators have two formats to discuss their observations – a synchronous video collaboration and an asynchronous text-based discussion board. The focus is on developing both the skill to identify the mathematics they consider evident in the preverbal children's actions and the willingness to share their observations with their peers and their educator. However, the interpretations of what is seen (Franzén, 2015, pp. 47) and what is believed to have been seen do not necessarily present the reality (Barad, 2007, p. 353).

This can be ameliorated through the discussion of what is observed, where all viewers thoroughly discuss their interpretations. As there may be concerns in terms of how what is observed on the video is categorised into Bishop's (1988, pp. 182–183) mathematical activities – Johansson et al. (2016, p. 33) highlighted the importance of identifying "who is doing the classification and for what purpose" – the reasoning behind choices needs to be fully considered.

Although this approach addresses the pre-service early childhood educators' situation-specific skill of noticing the mathematics in what preverbal children do, it also has the capacity to address the pre-service early childhood educators' mathematical self-efficacy and confidence (Cohrssen & Tayler, 2016, p. 35). Potential problems in terms of how the pre-service early childhood educators interact with their new knowledge and understandings of mathematics may not be positive – for those who fear mathematics and engaging with mathematics (Bates et al., 2013, p. 8), this may be overwhelming. To help the pre-service early childhood educators engage fully without fear of appearing unknowledgeable, they use pseudonyms (preventing identification from the educator and their peers); however, they may still feel hesitant in fully sharing their thoughts on the mathematics they have identified.

Conclusion

Bishop (1988, p. 180) describes mathematics as knowledge that is generated by all cultures, but which may look different from culture to culture. This is not surprising, as he considers mathematics as based on the context in which it is needed and therefore created (1988, p. 182). As young children are part of their culture and the context created by their culture (AGDET, 2009, p. 20), they will engage in mathematical thinking that is generated from these. Developing pre-service early childhood educators' skill to identify the mathematics in what preverbal young children do, and their using of this in their planning for future activities, will impact on the young children's learning (Meaney, 2016, p. 20). It is critical for pre-service educators to develop this skill as children are all different – even very young children may have engaged with different mathematical thinking and ideas due to their different environments and their efforts to make sense of those environments (Björklund, 2008, p. 88).

In becoming skilful at noticing and identifying mathematical thinking, pre-service early childhood educators will engage with the children in a more holistic manner. Their capacity to identify the mathematics in what the preverbal child does will enable them to attend to and engage with each child more thoroughly. As a result, the child will benefit from the focused attention, concentrating more fully and for longer (Björklund, 2012, p. 225) or extending their way of engaging with the phenomena of interest. The increased opportunities for young children to engage with mathematics is an aim reflected in Australia by the EYLF (AGDET, 2009, pp. 35, 38) and a demonstration of the educator's knowledge of children's development of numeracy (AITSL, 2011, p. 13).

References

Australian Government Department of Education and Training [AGDET] (2009). *Belonging, being, & becoming: The early years learning framework for Australia.* Retrieved from https://k10outline.scsa.wa.edu.au/home/teaching/early-years-learning-framework (accessed 07.02.2020).

Australian Institute for Teaching and School Leadership [AITSL] (2011). *Australian professional standards for teachers.* Carlton South, Australia: Education Services Australia. Retrieved from https://www.aitsl.edu.au/teach/standards (accessed 29.04.2020).

Barad, K. (2007). *Meeting the universe half way: Quantum physics of the entanglement of matter and meaning.* Duke Universal. https://doi.org/10.1215/9780822388128

Bates, A. B., Latham, N. I., & Kim, J. A. (2013). Do I have to teach math? Early childhood pre-service teachers' fears of teaching mathematics. *Issues in the Undergraduate Mathematics Preparation of School Teachers, 5.* Retrieved from http://files.eric.ed.gov/fulltext/EJ1061105.pdf (accessed 29.04.2020).

Bishop, A. J. (1988). Mathematics education in its cultural context. *Educational Studies in Mathematics 19*(2), 179–191. https://doi.org/10.1007/BF00751231

Bishop, A. (1991). *Mathematical enculturation: A cultural perspective on mathematics education* (Vol. 6). Springer Science & Business Media. https://doi.org/10.1007/978-94-009-2657-8

Björklund, C. (2008). Toddlers' opportunities to learn mathematics. *International Journal of Early Childhood, 40*(1), 81–95. https://doi.org/10.1007/BF03168365

Björklund, C. (2012). What counts when working with mathematics in a toddler-group. *Early Years. An International Journal of Research and Development, 32*(2), 215–228. https://doi.org/10.1080/09575146.2011.652940

Björklund, C. (2018). Powerful frameworks for conceptual understanding. In V. Kinnear, M. Lai, & T. Muir (Eds.), *Forging connections in early mathematics teaching and learning* (pp. 37–53). Singapore: Springer. https://doi.org/10.1007/978-981-10-7153-9

Björklund, C., & Pramling, N. (2017). Discerning and supporting the development of mathematical fundamentals in early years. In P. Sullivan, S. Phillipson, & A. Gervasoni (Eds.), *Engaging families as children's first mathematics educators: International perspectives* (pp. 65–80). Springer. https://doi.org/10.1007/978-981-10-2553_5

Blömeke, S., Gustafsson, J.-E., & Shavelson, R. J. (2015). Competence viewed as a continuum. *Zeitschrift für Psychologie, 223*(1), 3–13. https://doi.org/10.1027/2151-2604/a000194

Bronfenbrenner, U. (1977). Toward an experimental ecology of human development. *American Psychologist, 32*(7), 513–531. https://doi.org/10.1037/0003-066X.32.7.513

Cohrssen, C. & Tayler, C. (2016). Early childhood mathematics: A pilot study in preservice teacher education. *Journal of Early Childhood Teacher Education, 37*(1), 25–40. https://doi.org/10.1080/10901027.2015.1131208

Cooke, A. (2018). Mathematics in Swedish and Australian early childhood curricula. In E. Norén, H. Palmér, & A. Cooke (Eds.), *Nordic research in mathematics education.* Papers of NORMA 17 the eighth Nordic Conference on Mathematics Education Stockholm, May 30–June 2, 2017 (pp. 1–10). Stockholm, Sweden: Svensk Förening för MatematikDidaktisk Forskning. Retrieved from http://matematikdidaktik.org/wp-content/uploads/2018/09/NORMA-17-2018-papers-SMDF-skriftserie.pdf (accessed 29.04.2020).

de Freitas, E. (2016). The moving image in education research: Reassembling the body in the classroom video data. *International Journal of Qualitative Studies in Education, 29*(4), 553–572. https://doi.org/10.1080/09518398.2015.1077402

Franzén, K. (2015). Under threes' mathematical learning. *European Early Childhood Education Research Journal, 23*(1), 43–54. https://doi.org/10.1080/1350293X.2014.970855

Freudenthal, H. (1968). Why to teach mathematics so as to be useful. *Educational Studies in Mathematics, 1*, 3–8. https://doi.org/10.1007/BF00426224

Johansson, M. L., Lange, T., Meaney, T., Riesbeck, E., & Wernberg, A. (2016). What maths do children engage with in Swedish preschools? *Mathematics Teaching, 250*, 28–33.

Lynch, H., & Stanley, M. (2018). Beyond words: Using qualitative video methods for researching occupation with young children. *OTJR: Occupation, Participation and Health, 38*(1), 56–66. https://doi.org/10.1177/1539449217718504

Meaney, T. (2016). Locating learning of toddlers in the individual/society and mind/body divides. *Nordic Studies in Mathematics Education, 21*(4), 5–28.

Palmér, H., Henriksson, J., & Hussein, R. (2016). Integrating mathematical learning during caregiving routines: A study of toddlers in Swedish preschools. *Early Childhood Education Journal, 44*(1), 79–87. https://doi.org/10.1007/s10643-014-0669-y

Stites, M. L., & Brown, E. T. (2019). Observing mathematical learning experiences in preschool. *Early Child Development and Care*, 1–15. https://doi.org/10.1080/03004430.2019.1601089

van Oers, B. (2014). The roots of mathematising in young children's play. In U. Kortenkamp, B. Brandt, C. Benz, G. Krummheuer, S. Ladel, & R. Vogel (Eds.), *Early mathematics learning. Selected papers of the POEM 2012 conference* (pp. 111–123). https://doi.org/10.1007/978-1-4614-4678-1_8

Zippert, E. L., Eason, S. H., Marshall, S., & Ramani, G. B. (2019). Preschool children's math exploration during play with peers. *Journal of Applied Developmental Psychology, 65*, 101072. https://doi.org/10.1016/j.appdev.2019.101072

Part III

Performance

Chapter 9

Early childhood teachers' selection of subskills-related activities and instructional approaches to foster children's early number skills

Lara Pohle, Lars Jenßen, Katja Eilerts

Introduction

Recognizing and acknowledging the importance of early childhood (EC) education has led to the academic empowerment of pre-schools and resulted in a revised understanding of educational goals and approaches. This realignment contributed to the pivotal achievement of implementing early education programs in most industrialized countries (OECD, 2006). However, these programs vary greatly between countries due to differences in pedagogical traditions, country-specific conditions, and educational goals (Oberhuemer, 2005). Although different approaches may lead to controversies as well as different realizations of instructional practices, researchers generally agree that math activities are beneficial for children's mathematical development (de Haan, Elbers, & Leseman, 2014).

From a skills-related point of view, the decisive role of early number skills is constantly stressed due to their predictive power of later achievement in mathematics (Krajewski & Schneider, 2009; Nguyen et al., 2016). However, the implementation of mathematical ideas might vary considerably due to the different educational and pedagogical perspectives on which instructional practices are based. Therefore, it seems reasonable to map the various approaches that are used by EC teachers to introduce mathematical ideas. This is especially important because the selection of activities contributes substantially to the development of children's mathematical skills (Chien et al., 2010; de Haan et al., 2014).

Following the theoretical elaboration, this chapter examines (1) subskills-related activities (e.g., counting) German EC teachers choose to foster children's early number skills and (2) which instructional approaches (e.g., direct instruction) EC teachers opt for in order to present mathematical ideas.

Early number skills

Early number skills – sometimes also referred to as early numeracy skills (Aunio, 2019), core numerical skills (Aunio & Räsänen, 2016), or symbolic number sense (Jordan, Glutting, & Ramineni, 2010) – are undeniably important for children's mathematical development. Due to this, researchers came up with

DOI: 10.4324/9781003172529-9

various models describing their development and structure (e.g., Aunio & Räsänen, 2016; Clements, Sarama, & Liu, 2008; Krajewski & Schneider, 2009). Although differences between the models exist, researchers seem to agree upon the multifactorial nature of early number skills (Shanley, Clarke, Doabler, Kurtz-Nelson, & Fien, 2017). Against this backdrop, several subskills are frequently discussed and many of them have been examined in relation to children's later mathematics achievement (e.g., Aubrey, Godfrey, & Dahl, 2006; Krajewski & Schneider, 2009). In the following, a summary of the respective subskills is given (for a detailed overview, see Clements & Sarama, 2009):

Subitizing relates to the quick and ad-hoc recognition of quantities containing up to four elements. It is often referred to as the basis for mathematical understanding and the acquisition of mathematical principles that can be transferred to larger quantities (Gelman & Gallistel, 1978; Van den Heuvel-Panhuizen & Elia, 2020). In this context, quantity discrimination and magnitude comparison are noteworthy concepts which enable children to compare quantities (Krajewski & Schneider, 2009; Xenidou-Dervou, Molenaar, Ansari, van der Schoot, & van Lieshout, 2017).

Verbal and object counting are striking features of children's early number skills and a lot of research has been done on this matter (e.g., Rodrigues, 2019). Children learn to count at a very young age, but it is only later that they start perceiving number words as distinct entities or recite them in the correct order (Fuson, 1988). Children have to realize that number words are linked to quantities (Aunio & Räsänen, 2016; Dehaene, 1992; Krajewski & Schneider 2009) and internalize that the last number word they recite provides information on the cardinality. To achieve this, children have to be introduced to one-to-one correspondence, which means that each object is counted only once by assigning exactly one number word to one object. Additionally, children must process that the order of counting objects as well as their spatial arrangement are irrelevant for the result of counting (Gelman & Gallistel, 1978). Finally yet importantly, number symbols play a crucial role and verbal-symbol or symbol-verbal transitions are important for children's mathematical development and participation in everyday life (Aunio & Räsänen, 2016).

Children can only develop ability in *comparing and ordering* activities if they have internalized the aspects mentioned above. Building upon knowledge of the relationship between number words and quantities, children are then able to discriminate close number words and start to realize that number-word sequences represent increasing quantities (Krajewski & Schneider, 2009). This enables children to sequence numbers and decide whether, for example, 'eight' is larger or smaller than 'nine'. Additionally, it helps children to grasp the concept of ordinal numbers and to build a 'mental number line' (Clements & Sarama, 2009; Siegler & Lortie-Forgues, 2014). Furthermore, children get an insight into part-whole-relations when they realize that smaller quantities are contained within bigger quantities (Fuson, 1988; Krajewski & Schneider, 2009; Kullberg & Björklund, 2019).

Based on their prior knowledge, children later come to understand that quantities change when elements are added or taken away (Krajewski & Schneider, 2009; Kullberg & Björklund, 2019) providing the basis for *early addition and subtraction*. Apart from exploiting patterns (e.g., conceptual subitizing), finger and verbal counting, retrieval or derived combinations (e.g., doubling or halving), children tend to use *counting strategies* (e.g., counting-all, counting-on, counting-on-from-larger, counting-down, skip-counting-by-twos) or model problems in ways which make them more accessible (e.g., using concrete objects) (Clements & Sarama, 2009). Additionally, children get insights into commutativity, associativity, and inversion (Aunio & Räsänen, 2016).

Starting from their knowledge on part-whole relations, children learn about *number composition* meaning that different combinations result in the same number, for example, 6 can be represented by 2 + 4, 3 + 3, or 5 + 1 (Clements & Sarama, 2009; Krajewski & Schneider, 2009). *Place value* is a precondition to construct knowledge of numbers bigger than 10 (Clements & Sarama, 2009) and refers to the magnitude of a digit in relation to its place within a numeral (e.g., '2' in '25' means '20'). *Multidigit addition and subtraction* also draw on number composition, as numbers must be split into their place value magnitudes. This decomposition is used to solve addition and subtraction problems, for example, when children count by tens and ones (for detailed information, see Clements & Sarama, 2009).

As the above-mentioned facets are typically acquired before formal schooling starts (Jordan, Kaplan, Locuniak, & Ramineni, 2007), they can be considered as important aspects to establish connectivity between informal learning in preschools and formal learning at schools.

Instructional approaches in early childhood mathematics education

Among educational researchers, it is widely accepted that teachers' real-life behavior in classrooms affects children's development (e.g., Blömeke, Gustafsson, & Shavelson, 2015). In this context, quality is a construct that has been addressed in numerous studies (e.g., Von Suchodoletz, Fäsche, Gunzenhauser, & Hamre, 2014). However, the relationship between quality and children's achievement is often weak, which is why de Haan and colleagues (2014) follow other EC researchers (e.g., Chien et al., 2010) and conclude that, aside from quality, the activities provided by EC teachers are necessary to be examined. The importance of activities is also recognized by Salminen, Hännikäinen, Poikonen, and Rasku-Puttonen (2013, p. 128) who figure that '[t]eachers can produce good results and high-quality teaching through varying practices, and even highly similarly rated teachers can produce very different teaching practices'. Due to this, the question naturally arises of how exactly EC teachers approach content in mathematics, meaning, which instructional practices do they use to convey mathematical ideas?

EC education has a long tradition and various ideas and pedagogical approaches have shaped it over this time. For example, EC education in the USA experiences a 'push down' effect, where curriculum and related methods shift from primary school to pre-school in order to achieve children's school readiness (Rimm-Kaufman & Sandilos, 2017). Also referred to as 'schoolification', this procedure is a topic of intense debate (Broström, 2017) and a teacher-directed skills-based approach, where 'the teacher plays a central, directive role, and instruction focuses on developing discrete academic skills in a predetermined order' (Chiatovich & Stipek, 2016, p. 4), might best describe this situation. Contrary to this approach, Germany with its socio-pedagogical tradition emphasizes that EC education is no preponement of school (Hasselhorn & Kuger, 2014) and follows a child-centered approach (OECD, 2011). Here, children's learning is based on informal activities, such as play or everyday situations that serve as a starting point to introduce mathematical ideas (Anders & Rossbach, 2015; Lee, 2017). However, it is not necessary (and maybe not even helpful) to treat these approaches as dichotomies. Instead, the integration of contrasting perspectives might be fruitful for learners, especially against the backdrop of inconsistent research findings (for an overview, see Chiatovich & Stipek, 2016). Additionally, Chiatovich and Stipek (2016) found that teachers adhere to different instructional strategies and incorporate aspects of skills-based as well as child-centered approaches. This chapter therefore presents an excerpt of different approaches found across pre-schools and education systems.

EC teachers within socio-pedagogically oriented countries frequently resort to spontaneously occurring everyday activities in order to foster children's mathematical learning (Gasteiger, Bruns, Benz, Brunner, & Sprenger, 2020). Within Germany, this practice is also known as *situational approach* (Zimmer, 2007). The chance of this approach lies within teachers recognizing the 'teachable moment' (Ginsburg, Lee, & Boyd, 2008) in order to turn a spontaneously occurring real-life or play situation into a situation of mathematical meaning (Gasteiger et al., 2020; Lee, 2017). The downside of this approach might lie within EC teachers' situational perception regarding mathematical aspects, which is rather weak (Dunekacke, Jenßen, Eilerts, & Blömeke, 2016). The same applies for in-service EC teachers and their interpretation skills (Lee, 2017). In the end, these findings give reason to reconsider this approach, as it might not lead to the desired outcomes regarding children's mathematical learning.

Several researchers across different countries have acknowledged the importance of *play-based learning* for mathematics and made it subject of various studies. Generally speaking, play takes up a crucial part of childhood as it ties in with children's everyday experiences and builds on their interests. Even on its own, play provides children with various opportunities to learn and 'allows the expression of mathematical thinking' (McGrath, 2010, p. 25). Regarding children's early mathematics learning, research findings point to the assumption 'that playful learning activities may lead to greater learning than engaging in non-playful activities' (Scalise, Daubert, & Ramani, 2017, p. 562). Play has proven to be effective in order to seize knowledge concerning various mathematical ideas (for an overview, see Wickstrom, Pyle, & DeLuca, 2019). However, the nature of play is manifold and must be considered when interpreting research results. Although play is generally thought to be an informal approach to learning, Pyle and Danniels (2017) introduced a fine-grained view of play along a continuum. Drawing on the limitations that have been mentioned within the context of the situational approach, Anders and Rossbach (2015) found that EC teachers' ability to recognize mathematical aspects in play-based situations is improvable. This might limit children's possibilities to learn from (guided) play.

Although Ginsburg (2006, p. 159) clearly states the importance and benefits that come along with the above-mentioned approaches, he also states that a 'child-centered approach is not sufficient. The teacher must do more than seize upon the teachable moment that arises spontaneously'. Hence, researchers advocate curricula containing academic goals and content (Hachey, 2013). This view is particularly prominent in the USA where one of the main goals of pre-school education is preparing children for school (Ginsburg, 2006). One such skills-based approach to learning is the use of *training programs*, which are instructionally designed interventions aiming to support the development of children's cognitive skills, for example, in mathematics. Training programs are highly structured in regards to content and they are primarily teacher-directed. Thus, training programs present a form of direct instruction where 'discrete concepts are broken down into smaller units that are deliberately sequenced and explicitly taught' (Wickstrom et al., 2019, p. 288). Despite studies revealing that training programs do indeed have an effect on children's learning in mathematics (Clements & Sarama, 2007; Krajewski, Nieding, & Schneider, 2008), other studies show that play-based learning settings have an advantage over training programs (Stebler, Vogt, Wolf, Hauser, & Rechsteiner, 2013).

Although the above-mentioned approaches are primarily considered in terms of differences, they also share some similarities. Due to this overlap, it appears difficult to establish clear-cut categories. Building on the idea as proposed for play (Pyle & Danniels 2017), a continuum might be suitable to integrate different approaches.

Research questions

As outlined above, EC teachers' choice of activities contributes to the development of children's mathematical skills. Therefore, we aim to answer the following research questions:

1 *What* kind of subskills-related activities do EC teachers choose to foster children's early number skills?
2 *How* do EC teachers introduce the subskills-related activities to the children, meaning which instructional practices do they use?

Method

Throughout Berlin and Brandenburg, $n = 25$ EC teachers were observed while providing children with mathematical activities. All teachers were female and between 22 and 54 years old ($M = 33.45$; SD = 9.36). The teachers completed their job training within the last five years and either graduated from vocational schools (72.7%) or from universities (18.2%).

Approximately two months before observation, the teachers received the information to offer a mathematical learning activity considering aspects of early number skills. In this regard, the instruction contained examples like counting forward or comparing quantities. Beyond this, the teachers were explicitly requested to choose settings that comply with their daily routine in order to prevent any kind of classroom staging. Information on age-appropriate activities was not given, as mixed-age cohorts usually compose pre-school groups throughout Germany (OECD, 2016). According to the instruction handed out to the teachers, the activity was intended to take 10–15 minutes.

For observation, one rater or – if staff capacities allowed it – two raters were assigned per classroom and took notes on the teachers' performance including the mathematical content as well as the activities presented. In case two raters observed one teacher, protocols were compared after the observation and diverging ratings were compromised. All raters ($n = 10$) shared a math didactic background and were trained systematically before the observations started (Pohle, Hosoya, Loftfield, & Jenßen, 2019).

In order to answer the research questions, the occurrence of subskills-related (e.g., counting) and approach-related (e.g., play-based learning) variables were analyzed. Multiple selections were possible, meaning that one teacher could score on different categories (e.g., counting forward, one-to-one correspondence, and reading numbers). Additionally, teachers scored on an activity as soon as it was observed once.

Results

The mathematical learning situations provided by the EC teachers took between 10 and 45 minutes ($M = 19.76$; SD = 9.13). Group size varied between 2 and 14 ($M = 7.88$; SD = 2.99) and the children joining the activities were between 2 and 7 years old ($M = 4.32$; SD = .83). All children attended mixed-age groups. Table 9.1 gives an overview of EC teachers' individual choices of subskills-related activities and provides examples from the observations.

As can be seen from Table 9.1, all major categories of subskills-related activities were observed at least once across classrooms. However, there exist differences in the occurrence of single activities between EC teachers.

Table 9.2 gives an overview of teachers' different choices of instructional practices.

Table 9.1 Subskills-related activities provided by EC teachers

Number of EC teachers	Activity	Example
Counting, object counting, and number-quantity relationship		
24	Counting forward	Children are asked to count as far as they can
0	Counting backward	–
5	Counting on	Children are asked to count on from 6
1	Skip counting	One child is given eight nuts. The teacher asks the child to hand the nuts to four children and explains that every child gets two nuts. The child is asked to count by twos
24	Object counting	Children are asked to count the girls (boys, all children) of the group
6	One-to-one correspondence	Children roll a dice, count the dots, and are asked to collect the respective number of chestnuts
2	Number invariance/ conservation	The teacher builds two lines of rods, one below the other. One line contains six rods and one contains five. The distances between the rods in the line containing fewer elements are bigger than the distances in the line with six rods. The teacher asks: 'Which line has more rods?'
22	Linking number words to quantities	Each child gets a picture card showing a digit. The teacher asks the children to match the respective number of objects with the picture card. Afterward, the teacher asks: 'Your picture shows a nine and you already took one object. How many objects do you still have to take?'
Number symbols		
15	Reading numbers	The teacher shows picture cards with digits and asks: 'Which number is this?'
1	Writing numbers	The teacher asks one child to write down arithmetic problems $(1 + 2; 7 + 8)$
Comparing and ordering		
9	Comparing and ordering quantities	The teacher asks two children to compare the number of chestnuts in front of them: 'What do you think: who has more chestnuts?' 'In the beginning you had 20 chestnuts, now you have 24. Is 24 more than 20?'
10	Examining number-word sequences	Sorting picture cards with numbers in ascending order: 'Which number is the first one?' 'Which number comes before/after 4?'
Early addition and subtraction		
3	Operations (addition and subtraction)	Each child gets one stick. The sticks are striped (red and blue block stripes). The teacher asks the children to determine the number of red and blue stripes. Afterward, the teacher asks the children to add the red and blue stripes

(continued)

Table 9.1 (Continued)

Number of EC teachers	Activity	Example
Number composition and place value		
9	Composition and decomposition of quantities/numbers	The teacher puts picture cards showing different numbers of dots (1, 2, 3, …) in front of the children. Afterward, the teacher asks the children to figure out which cards compose 9 While speaking about the age of the children and the teacher, the teacher asks the children to figure out which numbers form a part of 28

Table 9.2 Instructional practices chosen by EC teachers

Number of EC teachers	Activity	Example
Group composition		
1	Individual work with teacher	Teacher turns to a single child while the rest of the children are occupied with non-mathematical activities
0	Individual work without teacher	–
1	Small groups with teacher	The teacher divided the group into subgroups and joined one after another
0	Small groups without teacher	–
24	Whole group with teacher	The teacher leads the activity by asking mathematical questions or giving instructions addressing the whole group
0	Whole-group without teacher	–
Teaching practices		
0	Working on stations	–
0	Experiment	–
11	Circle without play	Children and teacher sit in a circle while talking about mathematical ideas elicited and directed by the teacher. Sometimes, worksheets were handed out to the children
6	Circle with play	The children played board games like 'Obstgarten' or 'Ludo' while the teacher came up with mathematical stimuli. In another group, children rolled a dice and performed an exercise as often as indicated by the number of dots on the dice (motoric play)

The vast majority of activities were characterized by whole-group instruction and, regardless of group composition, teachers were always involved. Almost half of the activities took place within a morning circle where children and teachers sat together and talked about mathematical ideas. In the course of a few learning opportunities, games were played. Beyond this, board games and motoric play only occurred occasionally.

After the observation, 56% of EC teachers stated that they often conduct similar activities and 16% said that the activity is representative of their work. Only a minority of teachers stated that they rarely (4%) or only sometimes (20%) provide children with similar activities.

Discussion

This chapter qualitatively explored EC teachers' choice of subskills-related activities and instructional practices to foster children's early number skills.

1 What kind of subskills-related activities do EC teachers choose to foster children's early number skills?

Regarding subskills-related activities, the results show that although all main categories are reflected within teachers' performance, aspects concerned with counting, object counting, and number-quantity relationship are overrepresented in contrast to the remaining categories. However, also within this facet, imbalances exist. Regarding counting strategies, results reveal that counting forward and object counting were dominant, whereas other strategies like skip counting or counting on occurred only occasionally. Similar findings apply to other important principles, for example, to the gap between one-to-one correspondence or number invariance/conservation and linking number words to quantities.

One could argue that the latter encompasses the aforementioned principles, as they are necessary to build relationships between numbers and quantities. However, explicit and separate introduction to these mathematical ideas might be helpful, especially for children with less prior knowledge or children who, regarding their development of early number skills (e.g., Krajewski & Schneider, 2009), can be located on a lower level of competence. Yet, a meaningful and developmentally appropriate proceeding for all children is not easy to specify, at least if age is taken as reference. In Germany, mixed-age cohorts characterize the composition of pre-school groups (OECD, 2016), meaning that, within one group, children's age ranges from three to six years and sometimes even from one to six years (Rossbach, 2009). This fact is also reflected in our sample where children's age within groups ranges between two and seven years and therefore presents EC teachers with the challenge to implement ideas that are appropriate for all levels of development. Following this idea, EC teachers might have chosen certain subskills-related activities over others, because they might consider them developmentally appropriate for their group, irrespective of children's age.

Regarding early addition and subtraction, it needs to be added that its absence might be explained by the fact that, in Germany, these ideas are part of formal schooling and not explicitly referred to in the educational plans of the federal states of Berlin and Brandenburg (Ministerium für Bildung, Jugend und Sport des Landes Brandenburg, 2006; Senatsverwaltung für Bildung, Jugend und Wissenschaft, 2014). In contrast, the majority of EC teachers considered number symbols. Although symbols also belong to formal aspects of mathematical learning, they are part of children's everyday life (e.g., house numbers or bus numbers) and therefore might seem more natural to be discussed.

Another attempt to explain the existing imbalances in EC teachers' choice of subskills-based activities focuses on EC teachers' characteristics. For example, the underrepresentation of certain subskills-related activities might go back to EC teachers' lack of mathematical pedagogical content knowledge. This explanation corresponds to the rather minor role mathematics takes on during EC teachers' training in Germany (Janssen, 2010) and might explain why some teachers neglected certain subskills-related activities. Teachers simply might not be aware of the importance of ordering, for example, and/or ways to implement respective activities into their mathematical practice. In this regard, EC teachers' beliefs might also play a role. Future research should tie in with existing studies on EC teachers' beliefs (e.g., Chen, McCray, Adams, & Leow, 2014) and examine if some skills-based activities are perceived more positive than others or if teachers feel more confident about some ideas compared to others.

2   How do EC teachers introduce the subskills-related activities to the children, meaning which practices do they use?

Answering this question is not easy as teaching approaches and settings turn out to be different depending on the respective education system. Within our sample, two features are striking: First, teachers preferred whole-group settings, which might restrict children's active participation (de Haan et al., 2014). Second, irrespective of group composition, teachers were involved in every single activity. To some extent, this omnipresence might be traced back to the instruction teachers received prior to observation or the observation itself, which might have resulted in classroom staging. However, we asked the teachers to choose a setting that depicts their usual routine. The teachers, who after the observation mostly stated that they often or even always conduct similar activities, confirmed this. Against this backdrop, it is notable that play-based learning rarely occurred. Maybe teachers do not know about the possibilities play opens to introduce mathematical ideas, which is in line with EC teachers' limited capacity to recognize mathematical aspects in play-based situations (Anders & Rossbach, 2015). Jenßen, Thiel, Dunekacke, and Blömeke (2019) also speculate that EC teachers have difficulties perceiving situations in play as mathematical.

Overall, teacher-lead activities dominated throughout the observations. It seems that teachers chose an activity in order to foster a specific skill, for example, one-to-one correspondence or composition of quantities. A skills-based approach where the teacher directs and instructs children might best describe the results of our chapter (Chiatovich & Stipek, 2016). Although this might not be perfectly in line with what is understood to be a child-centered approach and appears to be contradictory to Germany's socio-pedagogical tradition, teacher-managed activities should not be condemned. Instead, it might be helpful to orientate oneself by integrating the practices onto a continuum as proposed earlier: The majority of settings chosen by EC teachers in our sample are not formal in a way that they resemble learning in schools, yet they are planned with the teacher being in charge. In our sample, the introduction of mathematical ideas is not grounded in spontaneously occurring everyday situations and, to a large extent, not in play either. However, this does not have to necessarily imply something negative, as research shows that children's mathematical development might benefit from teacher-managed activities (e.g., Chiatovich & Stipek, 2016; de Haan et al., 2014), and, again, that EC teachers' sensitivity toward mathematical aspects in play-based situations needs enhancement (Anders & Rossbach, 2015). At this point, Ginsburg's (2006) opinion to include practices beyond the situational approach gains in importance. However, to be fair, it has to be admitted that activities based on spontaneous situations seizing the 'teachable moment' were unlikely to be observed due to the design of this chapter and the instruction given prior to observation. Therefore, more research is needed mirroring typical days in EC institutions.

As the above-mentioned assumptions for the occurrence or absence of specific subskills-related activities and instructional practices are only speculations, they need further investigation. Considering EC teachers' beliefs and mathematical pedagogical content knowledge might be fruitful in order to draw substantial conclusions. This is supported by research findings suggesting a relationship between EC teachers' knowledge, emotions, perception skills, and planning abilities (Dunekacke et al., 2016; Jenßen et al., 2019).

This chapter contributes to research by disentangling mathematical activities and instructional approaches provided by EC teachers. Overall, results indicate considerable variability regarding teachers' choices as well as the time spent on mathematical activities. These findings are in line with other studies and might contribute to differences in children's mathematical development (Piasta, Yeager Pelatti, & Miller, 2014). Therefore, researchers are advised to consider possible relations between subskills-related activities, instructional practices, and children's mathematical achievement levels.

References

Anders, Y. & Rossbach, H.-G. (2015). Preschool teachers' sensitivity to mathematics in children's play: The influence of math-related school experiences, emotional attitudes and pedagogical beliefs. *Journal of Research in Childhood Education, 29*(3), 305–322.

Aubrey, C., Godfrey, R., & Dahl, S. (2006). Early mathematics development and later achievement: Further evidence. *Mathematics Education Research Journal, 18*(1), 27–46.

Aunio, P. (2019). Early numeracy skills learning and learning difficulties – Evidence-based assessment and interventions. *Mathematical Cognition and Learning, 5,* 195–214.

Aunio, P. & Räsänen, P. (2016). Core numerical skills for learning mathematics in children aged five to eight years – A working model for educators. *European Early Childhood Education Research Journal, 24*(5), 687–704.

Blömeke, S., Gustafsson, J.-E., & Shavelson, R. J. (2015). Beyond dichotomies: Competence viewed as continuum. *Zeitschrift für Psychologie, 223*(1), 3–13.

Broström, S. (2017). A dynamic learning concept in early years' education: A possible way to prevent schoolification. *International Journal of Early Years Education, 25*(1), 3–15.

Chen, J.-Q., McCray, J., Adams, M., & Leow, C. (2014). A survey study of early childhood teachers' beliefs and confidence about teaching early math. *Early Childhood Education Journal, 42,* 367–377.

Chiatovich, T. & Stipek, D. (2016). Instructional approaches in kindergarten. *The Elementary School Journal, 117*(1), 1–29.

Chien, N. C., Howes, C., Burchinal, M., Pianta, R. C., Ritchie, S., Bryant, D. M., Clifford, R. M., Early, D. M., & Barbarin O. A. (2010). Children's classroom engagement and school readiness gains in prekindergarten. *Child Development, 81,* 1534–1549.

Clements, D. H. & Sarama, J. (2007). Effects of a preschool mathematics curriculum: Summative research on the Building Blocks Project. *Journal for Research in Mathematics Education, 38*(2), 136–163.

Clements, D. H. & Sarama, J. (2009). *Learning and teaching early math: The learning trajectories approach.* New York, NY and London: Routledge.

Clements, D. H., Sarama, J., & Liu, X. (2008). Development of a measure of early mathematics achievement using the Rasch model: The research-based early maths assessment. *Educational Psychology, 28*(4), 457–482.

de Haan, A. K. E., Elbers, E., & Leseman, P. P. M. (2014). Teacher- and child-managed academic activities in preschool and kindergarten and their influence on children's gains in emergent academic skills. *Journal of Research in Childhood Education, 28*(1), 43–58.

Dehaene, S. (1992). Varieties of numerical abilities. *Cognition, 44*(1–2), 1–42.

Dunekacke, S., Jenßen, L., Eilerts, K., & Blömeke, S. (2016). Epistemological beliefs of prospective preschool teachers and their relation to knowledge, perception, and planning abilities in the field of mathematics: A process model. *ZDM Mathematics Education, 48,* 125–137.

Fuson, K. C. (1988). *Children's counting and concepts of number.* New York, NY: Springer.

Gasteiger, H., Bruns, J., Benz, C., Brunner, E., & Sprenger, P. (2020). Mathematical pedagogical content knowledge of early childhood teachers: A standardized situation-related measurement approach. *ZDM Mathematics Education, 52*(2), 193–205.

Gelman, R. & Gallistel, C. R. (1978). *The child's understanding of number.* Cambridge: Harvard University Press.

Ginsburg, H. P. (2006). Mathematical play and playful mathematics: A guide for early education. In D. G. Singer, R. Michnick Golinkoff, & K. Hirsh-Pasek (Eds.), *Play = learning: How play motivates and enhances children's cognitive and social-emotional growth* (pp. 145–166). New York, NY: Oxford University Press.

Ginsburg, H. P., Lee, J. S., & Boyd, J. S. (2008). Mathematics education for young children: What it is and how to promote it. *Social Policy Report, 22*(1), 1–24.

Hachey, A. C. (2013). The early childhood mathematics education revolution. *Early Education & Development, 24*(4), 419–430.

Hasselhorn, M. & Kuger, S. (2014). Wirksamkeit schulrelevanter Förderung in Kindertagesstätten. *Zeitschrift für Erziehungswissenschaft, 17*(S2), 299–314.

Janssen, R. (2010). *Die Ausbildung frühpädagogischer Fachkräfte an Berufsfachschulen und Fachschulen. Eine Analyse im Ländervergleich.* WiFF Expertise. München: DJI.

Jenßen, L., Thiel, O., Dunekacke, S., & Blömeke, S. (2019). Mathematikangst bei angehenden frühpädagogischen Fachkräften: Bedeutsam für professionelles Wissen und Wahrnehmung von mathematischen Inhalten im Kita-Alltag? *Journal für Mathematik-Didaktik, 41*(2), 301–327.

Jordan, N., Glutting, J., & Ramineni, C. (2010). The importance of number sense to mathematics achievement in first and third grades. *Learning and Individual Differences, 20*(2), 82–88.

Jordan, N., Kaplan, D., Locuniak, M. N., & Ramineni, C. (2007). Predicting first-grade math achievement from developmental number sense trajectories. *Learning Disabilities Research & Practice, 22*(1), 36–46.

Krajewski, K., Nieding, G., & Schneider, W. (2008). Kurz- und langfristige Effekte mathematischer Frühförderung im Kindergarten durch das Programm "Mengen, zählen, Zahlen". *Zeitschrift für Entwicklungspsychologie und Pädagogische Psychologie, 40*(3), 135–146.

Krajewski, K. & Schneider, W. (2009). Early development of quantity to number-word linkage as a precursor of mathematical school achievement and mathematical difficulties: Findings from a four-year longitudinal study. *Learning and Instruction, 19*(6), 513–526.

Kullberg, A. & Björklund, C. (2019). Preschoolers' different ways of structuring part-part-whole relations with finger patterns when solving an arithmetic task. *ZDM Mathematics Education, 52*, 767–778.

Lee, J. E. (2017). Preschool teachers' pedagogical content knowledge in mathematics. *International Journal of Early Childhood, 49*(2), 229–243.

McGrath, C. (2010). *Supporting early mathematical development: Practical approaches to play-based learning.* London and New York, NY: Routledge.

Ministerium für Bildung, Jugend und Sport des Landes Brandenburg (2006). *Grundsätze elementarer Bildung in Einrichtungen der Kindertagesbetreuung im Land Brandenburg.* From https://mbjs.brandenburg.de/media/lbm1.a.3973.de/Grundsaetze_elementa-rer_Bildung.pdf (accessed 15.10.2020).

Nguyen, T., Watts, T. W., Duncan, G., Clements, D., Sarama, J., Wolfe, C., & Spitler, M. E. (2016). Which preschool mathematics competencies are most predictive of fifth grade achievement? *Early Childhood Research Quarterly, 36*, 550–560.

Oberhuemer, P. (2005). International perspectives on early childhood curricula. *International Journal of Early Childhood, 37*(1), 27–37.

OECD (2006). *Starting strong II. Early childhood education and care.* Paris: OECD Publishing.

OECD (2011). *Starting strong III. A quality toolbox for early childhood education and care.* Paris: OECD Publishing.

OECD (2016). *Starting strong IV. Early childhood education and care data country note: Germany.* From http://www.oecd.org/edu/school/ECECDCN-Germany.pdf (accessed 15.10.2020).

Piasta, S. B., Yeager Pelatti, C., & Miller, H. L. (2014). Mathematics and science learning opportunities in preschool classrooms. *Early Education and Development, 25*(4), 445–468.

Pohle, L, Hosoya, G., Loftfield, C. & Jenßen, L. (2019). Indicators measuring preschool teachers' stimulation quality. *Zeitschrift für Pädagogik, 65*(4), 525–541.

Pyle, A. & Danniels, E. (2017). A continuum of play-based learning: The role of the teacher in play-based pedagogy and the fear of hijacking play. *Early Education and Development, 28*(3), 274–289.

Rimm-Kaufman, S. & Sandilos, L. (2017). School transition and school readiness: An outcome of early childhood development. *Encyclopedia on early childhood development.* From http://www.child-encyclopedia.com/school-readiness/according-experts/school-transition-and-school-readiness-outcome-early-childhood (accessed 15.10.2020).

Rodrigues, M. (2019). The development of number sense in preschool: The counting of objects. In M. Licardo, I. Simoes Dias (Eds.), *Contemporary themes in early childhood education and international educational modules* (pp. 73–86). Maribor: University of Maribor Press.

Rossbach, H.-G. (2009). The German educational system for children from 3 to 10 years old. In R. M. Clifford & G. M. Crawford (Eds.), *Beginning of school: U.S. policies in international perspective* (pp. 53–67). New York, NY: Teachers College Press.

Salminen, J., Hännikäinen, M., Poikonen, P.-L., & Rasku-Puttonen (2013). A descriptive case analysis of instructional teaching practices in Finnish preschool classrooms. *Journal of Research in Childhood Education, 27*(2), 127–152.

Scalise, N. R., Daubert, E. N., & Ramani, G. B. (2017). Narrowing the early mathematics gap: A play-based intervention to promote low-Income preschoolers' number skills. *Journal of Numerical Cognition, 3*(3), 559–581.

Senatsverwaltung für Bildung, Jugend und Wissenschaft (2014). *Berliner Bildungsprogramm für Kitas und Kindertagespflege.* Weimar & Berlin: Verlag das Netz.

Shanley, L., Clarke, B., Doabler, C. T., Kurtz-Nelson, E., & Fien, H. (2017). Early number skills gains and mathematics achievement: Intervening to establish successful early mathematics trajectories. *The Journal of Special Education, 51*(3), 177–188.

Siegler, R. S. & Lortie-Forgues, H. (2014). An integrative theory of numerical development. *Child Development Perspectives, 8*(3), 144–150.

Stebler, R., Vogt, F., Wolf, I., Hauser, B., & Rechsteiner, K. (2013). Play-based mathematics in kindergarten: A video analysis of children's mathematical behaviour while playing a board game in small groups. *Journal für Mathematik-Didaktik, 34,* 149–175.

Van den Heuvel-Panhuizen, M. & Elia, I. (2020). Mapping kindergarteners' quantitative competence. *ZDM Mathematics Education, 52*(4), 805–819.

Von Suchodoletz, A., Fäsche, A., Gunzenhauser, C., & Hamre, B. K. (2014). A typical morning in preschool: Observations of teacher-child interactions in German preschools. *Early Childhood Research Quarterly, 29*(4), 509–519.

Wickstrom, H., Pyle, A., & DeLuca, C. (2019). Does theory translate into practice? An observational study of current mathematics pedagogies in play-based kindergarten. *Early Childhood Education Journal, 47,* 287–295.

Xenidou-Dervou, I., Molenaar, D., Ansari, D., van der Schoot, M., & van Lieshout, E. C. D. M. (2017). Nonsymbolic and symbolic magnitude comparison skills as longitudinal predictors of mathematical achievement. *Learning and Instruction, 50,* 1–13.

Zimmer, J. (2007). *Das kleine Handbuch zum Situationsansatz.* Weinheim: Beltz.

Chapter 10

Longitudinal evaluation of a scale-up model for professional development in early mathematics

Julie Sarama, Douglas H. Clements,
Shannon Stark Guss

Professional development in early mathematics: learning trajectories and tools

Teaching is a complex enterprise, and teaching early mathematics is particularly complex (National Research Council, 2009). Unfortunately, most early childhood teachers receive at best meager training in domains other than literacy (Institute of Medicine (IOM) and National Research Council (NRC), 2015). Thus, the field needs effective programs and tools for pre-service education and in-service professional development (PD), especially for scalable efforts. This chapter describes a PD program and a related resource based on children's learning trajectories that multiple evaluations have indicated can provide coherent and effective PD.

Theoretical background

Math is increasingly important in a modern global economy, but mathematics achievement in many countries has not matched this importance. Our own country, the USA, has fewer high-performing and more low-performing students than do high-performing countries, especially in math (http://ncee.org/pisa-2018-lessons/). These differences appear as early as first grade, kindergarten, and even pre-school (e.g., Gerofsky, 2015; Mullis, Martin, Foy, & Arora, 2012; OECD, 2014). Although those higher performing countries continue showing improvements, many like the USA are not (Mullis et al., 2012). This lack of progress is one reason interest in improving early childhood mathematics education has increased worldwide, often focusing on children who have not been provided opportunities to learn (McCoy et al., 2018).

A central reason for limited learning is that many early childhood teachers are not well prepared to engage children in rich experiences in mathematics (Institute of Medicine (IOM) and National Research Council (NRC), 2015; National Research Council, 2009). Fortunately, research provides some guidance regarding the essential features of high-quality PD (Björklund & Barendregt, 2016; Brendefur, Strother, Thiede, Lane, & Surges-Prokop, 2013; Carpenter, Fennema, Franke, Levi, & Empson, 2014; MacDonald, Davies, Dockett, & Perry, 2012; Perry, 2010; Piasta, Logan, Pelatti, Capps, & Petrill, 2015; Sarama & Clements, 2019;

DOI: 10.4324/9781003172529-10

Sarama, Clements, Wolfe, & Spitler, 2016). In developing our approach, we followed this guidance, placing a unique emphasis on *learning trajectories.*

We emphasize learning trajectories because they support, by definition, the type of PD that guides development of teachers' pedagogical *competence* based on specific *competencies* (Blömeke, Gustafsson, & Shavelson, 2015). We first briefly describe these competencies, then provide our definition of learning trajectories, and then relate these two.

The first competency is understanding the goals of mathematics education, which involves two related ideas: standards and content. Teachers need to understand what is expected of children at a given age (the standards[1]). However, to fully understand those and all other aspects of teaching, teachers need what Shulman (1986) called content knowledge—knowledge of the structure, syntax, and even epistemology of the domain; "not just understanding *that* something is so, but also *why* it is so."

The second competency is understanding children's thinking and learning in mathematics. Especially in early childhood, it is necessary for teachers to take children's perspective, as their ideas about mathematics can be fundamentally different from those of adults, who often cannot recall early ideas, such as non-conservation. This is one part of Shulman's (1986) pedagogical content knowledge—the conceptions and preconceptions that children of different ages and backgrounds bring with them to the learning situation. Most researchers in PD agree that teachers need to understand children's mathematical thinking (e.g., Baroody & Ginsburg, 1990; Franke, Carpenter, Levi, & Fennema, 2001), with documented connection between such understanding and achievement (e.g., Carpenter, Franke, Jacobs, Fennema, & Empson, 1998). Children's thinking does not always follow historical or logical developments of mathematics as a discipline, nor adult reasoning (Björklund & Barendregt, 2016; Sztajn, Confrey, Wilson, & Edgington, 2012). However, it follows predictable sequences of levels, or patterns, of thinking (Battista, 2013; Bobis et al., 2005; Clarke, 2008; Confrey, Maloney, Nguyen, & Rupp, 2014; Sarama & Clements, 2009). Teachers do not assume children's thinking is a proper subset of adults' knowledge, but instead, they learn how to understand mathematics as children think and learn it.

The third competency is understanding effective teaching. This broad category includes several of Shulman's types of knowledge. They include knowledge of effective forms of representing and presenting content, curricular knowledge, awareness of instructional materials, and effectiveness.

We turn to our definition of learning trajectories.

> We conceptualize learning trajectories as descriptions of children's thinking and learning in a specific mathematical domain and a related, conjectured route through a set of instructional tasks designed to engender those mental processes or actions hypothesized to move children through a developmental progression of levels of thinking, created with the intent of supporting children's achievement of specific goals in that mathematical domain.
>
> (Clements & Sarama, 2004, p. 83)

In other words, each learning trajectory has three components: (a) a goal, (b) a developmental progression, and (c) instructional activities. To attain a mathematical competence in a given topic or domain (the goal), students learn each successive level (the developmental progression), aided by tasks (instructional activities) designed to build the mental actions-on-objects that enable thinking at each higher level.

These three components of LTs, then, map well onto the three pedagogical competencies. To use LTs, then, teachers have to understand the goal—the mathematics as subject-matter content and the mathematics to be learned. Second, they need to understand the developmental progression—observing and identifying children's mathematical thinking. Third, teachers must understand the instructional tasks, not merely how to enact instructional activities, but *when* to implement them, *with whom*, and *why*. Learning trajectories-based PD goes beyond teaching specific competencies to teachers but develops sustainable *pedagogical competence* (Blömeke et al., 2015) because the learning trajectory *framework dynamically connects all three components*—mutually reinforcing learning through practice.

Before elaborating on these ideas, we wish to recognize that teaching involves additional knowledge and skills. Shulman's (1986) general pedagogical knowledge is also needed. So are affect-motivation underpinnings (Blömeke et al., 2015), although our theory and research suggest that beliefs and motivation/commitment may follow successful new practices. Here, we focus on developing pedagogical competence in the teaching of early mathematics.

The succeeding sections describe our development and evaluation of learning-trajectories-based PD, including the development of online tools that were fundamental to the PD and especially our effects of scaling up and widely disseminate resources for others to use in their own PD efforts.

Development of the PD model

Because little was known about effective PD in early mathematics, one of us conducted a National Science Foundation planning grant. Results of that project (Sarama, 2002) and extensive reviews of research undergirded our PD model. For example, support for basing PD on LTs comes from studies that show learning trajectories-based instruction served as a tool for teacher education by providing a framework for organizing theory on teaching and learning (Sztajn et al., 2012). For example, an LT on equi-partitioning supported teachers' understanding of mathematics, as well as their students' reasoning (Wilson, Mojica, & Confrey, 2013). This approach to PD was supported by the Association of Mathematics Teacher Educators Standards for Mathematics Teacher Preparation (Association of Mathematics Teacher Educators, 2017) that emphasized the central role learning trajectories should play, especially for early childhood.

Definition of the model

In this section, we define our PD model and then describe how we enacted it, including the development of our online tool. The following is our description of the approach (Clements & Sarama, 2008; Sarama & Clements, 2019).

Provide PD that is ongoing, intentional, reflective, goal-oriented focused on content knowledge and children's thinking, grounded in particular curriculum materials, and situated in the classroom and the school. A focus on content includes accurate and adequate subject-matter knowledge both for teachers and for children. A focus on children's thinking emphasizes the learning trajectories' developmental progressions and their pedagogical application in formative assessment. Grounding in particular curriculum materials should include all three aspects of learning trajectories, especially their connections. This also provides a common language for teachers in working with each other and collaborating groups (Bryk, Sebring, Allensworth, Suppescu, & Easton, 2010). Situated in the classroom does not imply that all training occurs within classrooms. However, off-site intensive training remains focused on and connected to classroom practice and is completed by classroom-based enactment with coaching. In addition, this PD should encourage sharing, risk-taking, and learning from and with peers (e.g., Bryk et al., 2010; Elmore, 1996; Guskey, 2000; Hall & Hord, 2001; Pellegrino, 2007; Showers, Joyce, & Bennett, 1987).

Enactment of the model

We followed these research guidelines in multiple implementations (Clements & Sarama, 2008; Clements, Sarama, Spitler, Lange, & Wolfe, 2011; Sarama, Clements, Starkey, Klein, & Wakeley, 2008), recognizing their importance for implementation with fidelity at any scale but also that the planning, structures, common language, formative evaluation, and school-level context are increasingly crucial as the implementation moves to larger scales. We refined the approach with each implementation; here, we describe the latest implementation at the largest scale.

To ensure the PD was ongoing, we planned eight full days for the first year, and five for the second, with coaching one to two times per month. We grounded it in our pre-school mathematics curriculum based on learning trajectories, thus emphasizing children's thinking.

Indeed, all three components of the learning trajectories were emphasized. The first, the mathematical goal, involved brief presentations on mathematical concepts and ideas, followed by extensive work on mathematical explorations and problems (at the pre-K level and above), to facilitate teachers' understanding of the mathematical concepts, practices, and skills for each main topic (Polly et al., 2015; Stosich, 2016), as well as how they connected specific educational objectives (NCTM, 2006; NGA/CCSSO, 2010). For example, work in geometry included the pre-K activities, of course, but far beyond these, such as placing a diverse set of shapes in Venn diagrams, challenging each other with "guess my rule" games, and exploring tilings of pattern blocks and tangrams, with extension discussions about mathematical justifications for ideas and hypotheses these activities engendered. In an illustration of geometric measurement, teachers were given an 8.5 by 11 cm picture and asked to figure out what number would scale it (e.g., on a copy machine) to fit it to an 8.5 by 11-inch sheet of paper.

The second component of the learning trajectory was introduced with a newly developed online tool, described in detail in the following section. Here, teachers were shown interviews of children who were at different levels of the developmental progression. Teachers described the behaviors and their interpretation of children's thinking that might underlie the behaviors. Focus was on the qualitatively different patterns of thinking between contiguous levels. They then used the online tool to study and analyze new videos of children's thinking through the developmental progression (see the following section).

The third component of the learning trajectory was introduced with the teachers' edition of the printed curriculum and videos from the online tool. Activities were introduced and briefly modeled in plenary sessions. Essential for teachers practicing activities were two characteristics: (a) the room was set up to mirror a pre-K classroom, with whole group, small group, and centers (including a computer center); and (b) in each of these contexts, teachers read the teachers' guide before enacting each activity—without the facilitator modeling all steps of the activity. Because teachers can often duplicate actions without interpreting the text, the facilitator assured them that reading and taking turns implementing activities in the safe space of the PD would enhance the enactment of activities in their own classrooms. Thus, one group of teachers gathered for whole-group activities in a simulated setting, enacting the whole group lessons by taking turns reading and leading the lesson, with others playing the part of pre-schoolers. They similarly practiced small-group activities by gathering in groups of five to simulate the small group sessions they would lead in their classrooms. At the computer, teachers also independently practiced the curricular computer activities they would introduce to the children in the upcoming weeks. Additionally, teachers could interact with the data management system that tracked the progress of each of their children on every software activity. Tracking children's progress would then allow teachers to tailor assigned software activities to the level of most children in the class and to individuals (i.e., if any children needed either earlier or more challenging activities) at the beginning of each week.

Any time teachers had questions, and also during dedicated time later in the day, they used the online tool to view videos of instructional activities implemented by multiple teachers. The tool *linked* these to the level of thinking they promoted. This raises an essential aspect of the PD regarding learning trajectories: Each of the three components was addressed, but *connections were emphasized throughout the PD*. For example, formative assessment of children's place on the continuum of the learning progressions continuously emphasized the need to understand children's thinking. It was an expected part of the small-group implementation. Teachers became familiar with the record sheets they would use to formatively assess every student during every small-group lesson they taught in their classrooms. This became routine for the teachers over time. They became skilled at using the names of the developmental progression's LT levels (e.g., for counting beyond verbal counting, "corresponder," "counter" (small numbers), and "producer"). Knowledge of these terms facilitated quick and accurate formative assessments. Teachers also discussed how to observe, listen to, and interpret children's thinking. They learned that formative assessment is a powerful tool for decision-making, for knowing what to teach to whom and when, and for considering a range of levels for children who were gifted and/or had special needs. *Then, teachers discussed the instructional activities for each level, analyzing the key characteristics that would promote children's learning of each level of thinking.* Thus, teachers needed to simultaneously consider the mathematical goal and content, developmental progression, instructional activities, and their interrelationships.

In the second year, teachers presented and discussed case studies with their peers, focusing on the mathematical learning trajectories and how they were used or adapted for groups or individual children. The common language for working with each other helped take their application of learning trajectories to a more proactive level. The ideas they shared for productive adaptations of the activities were assembled and published for everyone.

Consistent with the research-based model, PD also took place in the teachers' own classrooms via visits from project coaches and peer coaches. Project coaches were TRIAD staff who received two days of training on coaching early mathematics. They participated in all PD sessions and provided individualized support during these sessions. They also interacted with their assigned teachers (four to five teachers per coach) approximately twice per month. At times, the teacher would observe as the project coach demonstrated small group, whole group, and software activities, but usually, the project coach would observe and support while the teacher implemented the curricular activities. Project coaches completed fidelity ratings three to four times per year, provided teachers with feedback, and offered teachers a fidelity rating sheet that teachers could voluntarily use to self-evaluate.

School-based peer coaches were teacher participants who volunteered to lead all participating teachers at their school. They served as a liaison to the project staff and provided in-school/in-classroom support for the teachers as they implemented the intervention. Technology mentors (project staff) rotated into classrooms to troubleshoot software, hardware, or network problems, demonstrate software, and answer teachers' technology questions.

In summary, the last implementation of the research guidelines involved multiple components, all designed to support teachers in internalizing and applying knowledge of research-based learning trajectories.

Development and use of the online PD tool: BBLT

The original online PD tool was "Building Blocks Learning Trajectories" (BBLT), designed to help teachers learning the LTs as implemented explicitly in the building blocks curriculum. BBLT presented and connected the components of each learning trajectory, encouraging teachers to view them through a curriculum or developmental (children's thinking) perspective. Each view was linked to the other. That is, teachers might choose the instruction view (Figure 10.1a), then click an activity and not only see an explanation and video of the activity "in action," but also immediately see the level of thinking that activity is designed to develop, in context of the entire LT. For example, in Figure 10.1a, the user has selected Week 24. The activities are listed by type and a suggested weekly schedule is provided. The user reads the description that appears on the right. If they choose "More info," the screen "slides over" to reveal the expanded view shown in Figure 10.1b. Here they can see multiple video examples with commentary. Clicking on the related developmental level (child's level of thinking) yields the view in Figure 10.1c. This developmental view likewise provides a description, video, and commentary on the developmental level—the video here is of a clinical interview task in which a child displays that level of thinking.

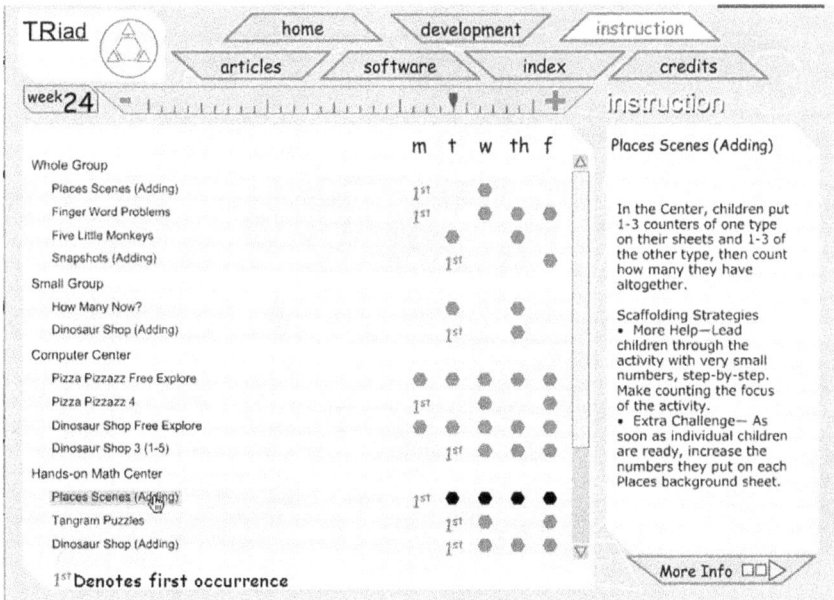

Figure 10.1a The building blocks learning trajectories (BBLT) online tool.

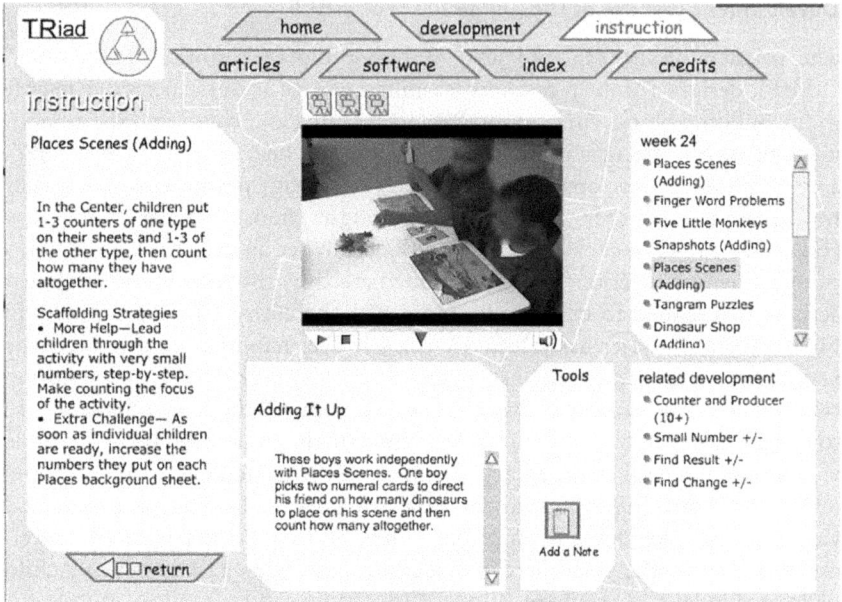

Figure 10.1b The building blocks learning trajectories (BBLT) online tool.

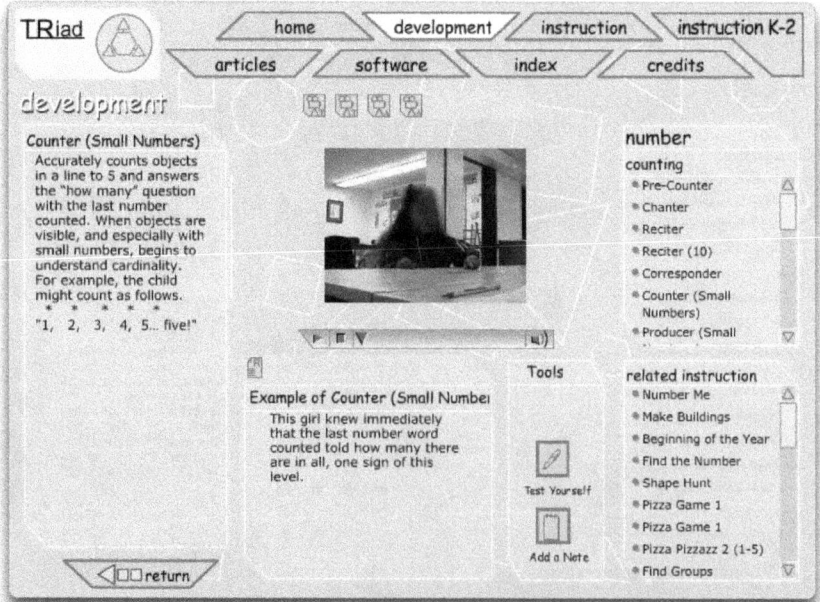

Figure 10.1c The building blocks learning trajectories (BBLT) online tool.

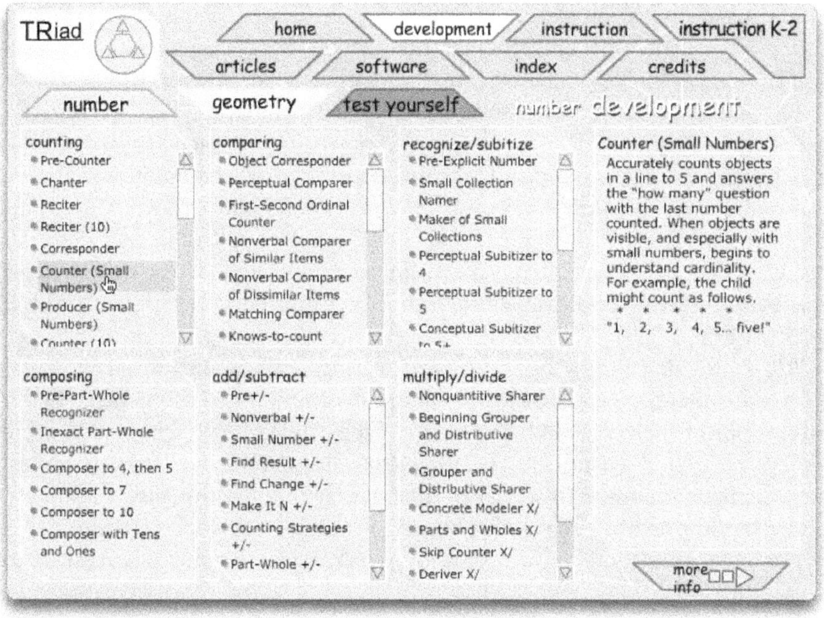

Figure 10.2 BBLT viewed through the developmental progression.

Alternatively, the user may have been studying developmental sequences themselves. After viewing the complete list of the number topics on the left (Figure 10.2), she may have selected the same level of the counting LT shown in Figure 10.1c.

Pressing "More info" results in the same developmental view as Figure 10.1c. The video commentary shown is just one of three commentaries, some by researchers, some by assessors, and still others by teachers. Further, the level is illustrated with assessment tasks and video of classroom activities in which children illustrate thinking at each level, an approach that has received empirical support (Klingner, Ahwee, Pilonieta, & Menendez, 2003).

Finally, this developmental level was connected to *all* correlated activities. Thus, a user in this view could jump to the "Make Buildings" activity *or* any of the activities whose goal is to develop that level of thinking. This allowed the teacher to use activities that may be needed due to children needing more time to learn the level or to match children's interests.

Teachers also could test themselves over the developmental progressions component of the LTs by seeing a video and attempting to classify the level of thinking displayed with assistive feedback is offered on their attempts. This further solidified their learning about and confidence in their understanding of children's mathematical thinking.

The same videos could serve multiple purposes—an example of a curriculum activity when studying developmental sequences or an example of a particular child's level of thinking. In each case, the video's supporting text directed attention to each perspective and the connections between them. Such rich learning experiences promoted flexible, integrated knowledge, as teachers learned to see and integrate the teaching and learning aspects of education. Research confirms that such learning is necessary to successfully apply concrete cases and the theories they are embedded in (Feltovich, Spiro, & Coulson, 1997). The resulting cognitive flexibility positively impacts the variety of teaching strategies that people develop and ease with which they acquire new repertoires (Showers et al., 1987).

BBLT provided PD by bringing participating teachers into intimate contact with "best practice" classrooms, including instruction and assessment. Online discussions of the videos of best practice made explicit how such practice exemplified research-based principles. Discussions also emphasized that even exemplary teachers continue to struggle—illustrating that high-quality teaching for understanding is both rewarding and challenging—everyone can continue to improve and contribute to the profession (Heck, Weiss, Boyd, & Howard, 2002; Weiss, 2002).

Even with these experiences, "most teachers report that they believe their children are capable of fine work, but what they think they know from daily experience often hedges that belief with limited expectations" (Ball & Cohen, 1999, p. 8). This leads to another feature of *BBLT*. Teachers have an opportunity to assess and teach *their* children with our curriculum and post their experiences on *BBLT*—including their own narrative (and, for some, photographs and even short video segments), to seek reactions from and further discuss shared issues with others.

BBLT is used in four related ways. First, it aids presentations of the trajectories and activities to teachers. Second, teachers observe, react to, test themselves on, and discuss (often online) specific trajectory levels, activities, or the relationship between them. Third, coaches and mentors use the site in conversations with teachers, often in their classrooms, about the trajectories, activities, or their relationship. This is especially valuable in situations in which a teacher says or demonstrates that they did not fully understand a given activity's goals or structure. Fourth, teachers may voluntarily consult BBLT when they wish to refresh their memories on a particular activity or delve more deeply into understanding their children's thinking.

Evolution of the online tool: Learning and Teaching with Learning Trajectories—[LT]²

At the time of this writing, BBLT is off-line because the code was on an unsupported software platform. Recognizing that this would eventually happen, we have planned for years to replace BBLT. In so doing, we wished to shore up several limitations in BBLT by implementing the following ten improvements.

1   Move to more platforms (e.g., phones and tablets), simultaneously upgrading and improving the interface. In contrast, BBLT was limited to computers and used software not available on many platforms.
2   Upgrade the video from BBLT's older, lower resolution video.
3   Expand the range of the learning trajectories for all children birth to grade 3 from BBLT's primary focus on pre-K.
4   Enhance instructional activities, replacing most of those from a copyrighted curriculum, as BBLT was available only to those owning the Building Blocks teachers' guide due to copyright limitations.
5   Collect anonymized data on users' use and navigation of the program.
6   Support use by Spanish speakers.

We implemented these improvements with funding from the Heising-Simons Foundation and the Bill & Melinda Gates Foundation.

Thus, the new online tool, Learning and Teaching with Learning Trajectories (LTLT, or [LT]²), website at www.LearningTrajectories.org is a PD tool for teachers and caregivers to support early math instruction from birth to grade 3 (Figure 10.3a).

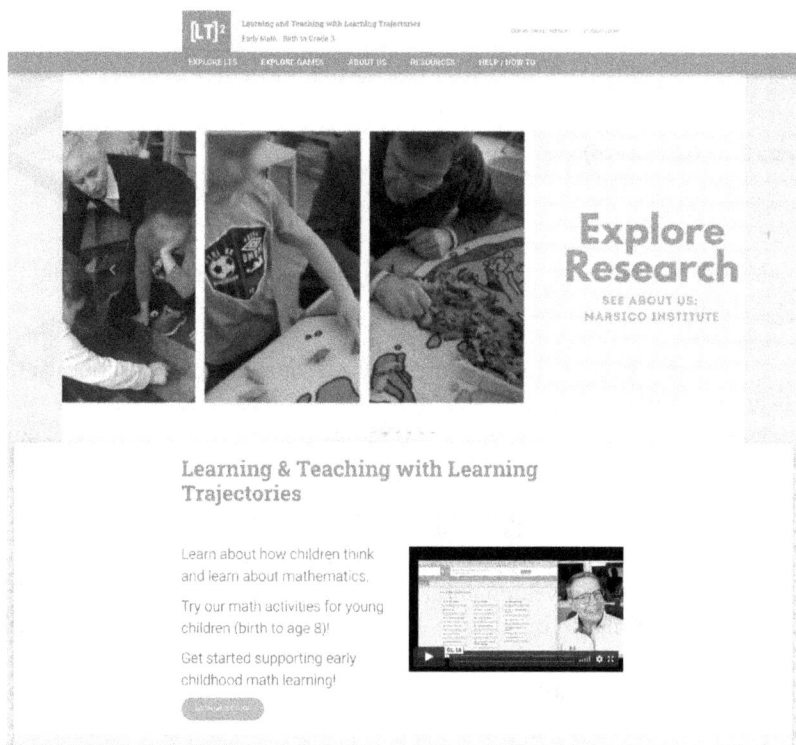

Figure 10.3a The Learning and Teaching with Learning Trajectories (LTLT or [LT]²) tool at www.LearningTrajectories.org.

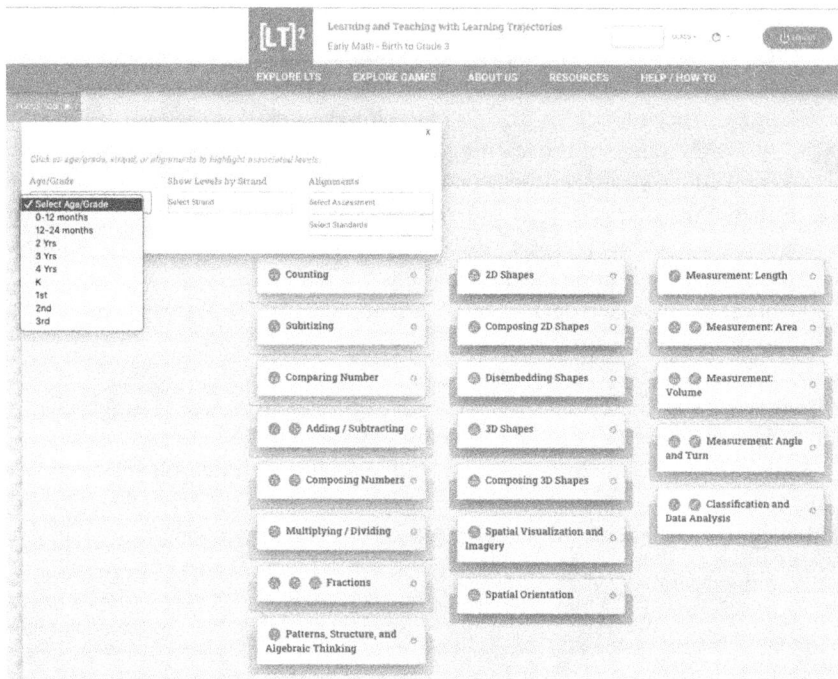

Figure 10.3b (Continued).

In addition to those planned and listed above, [LT]² includes additional improvements.

1 [LT]² covers far more topics than BBLT and shows alignments for each one (Figure 10.3b).
2 Based on recent research, [LT]² includes revised levels for each LT (Figure 10.3c).
3 Further, [LT]² includes a new section, "Learn About…," that addresses the first component of all LTs, the *goal* (Figure 10.3c).
4 Videos illustrating particular characteristics of children's thinking at each level of the developmental progressions were carefully selected to serve as landmarks or growth points (Clarke et al., 2001) for teachers to identify levels of their own children's mathematical thinking, as well as children's growth through the developmental progression (top of Figure 10.3d).
5 Each level of the developmental progression is linked to instructional activities on the same page, which are similarly illustrated by videos and pictures. This provides more opportunities to see connections between

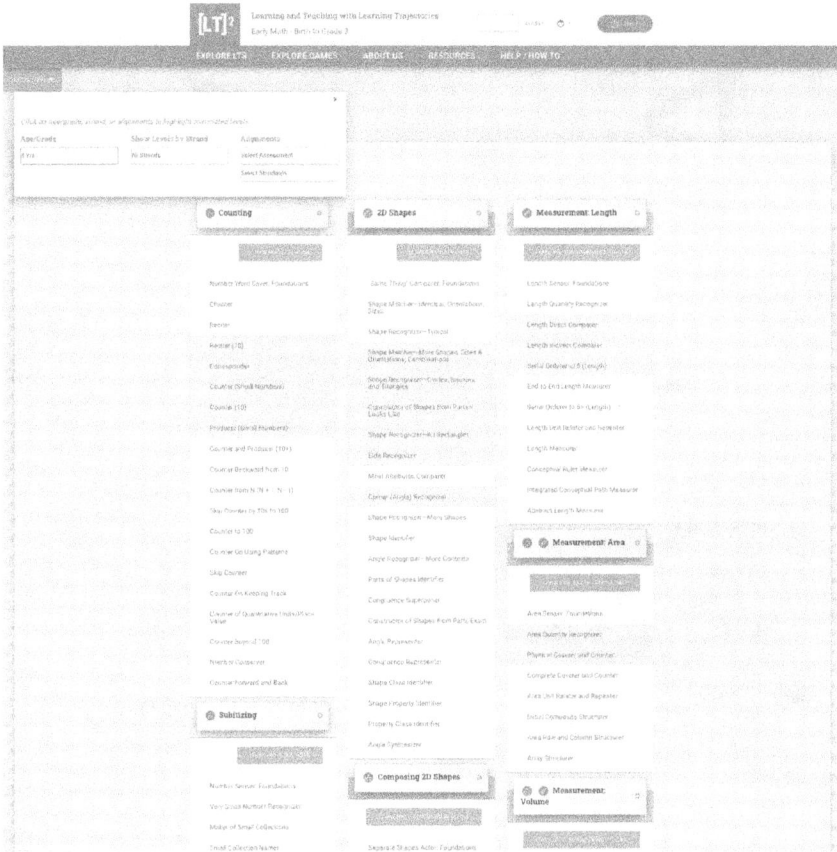

Figure 10.3c (Continued).

the learning trajectory's level of thinking and the specific instructional activities designed to promote children's learning of that mathematical competency.

6 A paragraph introducing the activities describes how the activities were designed and why the design will support that level of thinking (Figure 10.3d).

7 Clicking on one of the activities (in Figure 10.3d) takes you to a new page for that activity, offering video instructions and a set of downloadable, carefully formatted PDF files of the activity in English and Spanish, as well as printable materials for the activities when applicable (Figure 10.3e).

8 A resources page supports more users, including teachers, home visitors, caregivers and parents, and guides for PD, while BBLT was mainly a tool for those providing PD (Figure 10.3f).

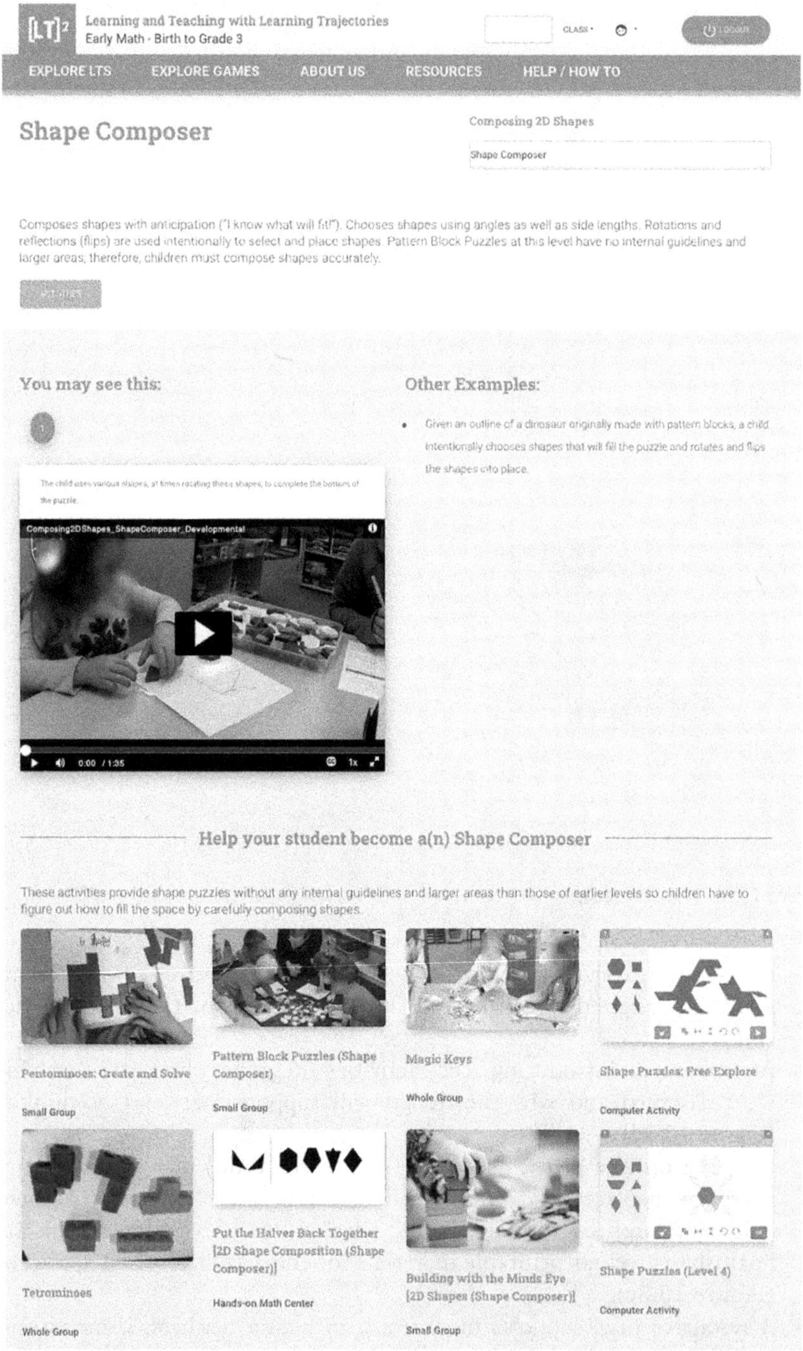

[LT]² Learning and Teaching with Learning Trajectories
Early Math - Birth to Grade 3

CLASS ·   ·   LOGIN

EXPLORE LTS   EXPLORE GAMES   ABOUT US   RESOURCES   HELP / HOW TO

Shape Composer

Composing 2D Shapes

Shape Composer

Composes shapes with anticipation ("I know what will fit!"). Chooses shapes using angles as well as side lengths. Rotations and reflections (flips) are used intentionally to select and place shapes. Pattern Block Puzzles at this level have no internal guidelines and larger areas; therefore, children must compose shapes accurately.

You may see this:

The child uses various shapes, at times rotating these shapes, to complete the bottom of the puzzle.

Composing2DShapes_ShapeComposer_Developmental

Other Examples:

- Given an outline of a dinosaur originally made with pattern blocks, a child intentionally chooses shapes that will fill the puzzle and rotates and flips the shapes into place.

——————— Help your student become a(n) Shape Composer ———————

These activities provide shape puzzles without any internal guidelines and larger areas than those of earlier levels so children have to figure out how to fill the space by carefully composing shapes.

Pentominoes: Create and Solve

Small Group

Pattern Block Puzzles (Shape Composer)

Small Group

Magic Keys

Whole Group

Shape Puzzles: Free Explore

Computer Activity

Tetrominoes

Whole Group

Put the Halves Back Together [2D Shape Composition (Shape Composer)]

Hands-on Math Center

Building with the Minds Eye [2D Shapes (Shape Composer)]

Small Group

Shape Puzzles (Level 4)

Computer Activity

Figure 10.3d (Continued).

[LT]² Learning and Teaching with Learning Trajectories
Early Math - Birth to Grade 3

CLASS •   ○ •   LOGOUT

EXPLORE LTS   EXPLORE GAMES   ABOUT US   RESOURCES   HELP / HOW TO

Pattern Block Puzzles (Shape Composer)

ACTIVITY TYPE: SMALL GROUP

Quick Description

Children use shapes to fill in a puzzle design. (Adapted From: Building Blocks)

You may see this:

The children are using pattern blocks to complete the puzzles. Notice that the children are working on puzzles that fit at different levels. The puzzles appropriate for this level do not have internal questions.

Composing2DShapes_ShapeComposer_PatternBlockPuzzles ❶

◀)) 0:00 / 3:02   ⚙ 1x

Directions:

- Provide each child with a puzzle design.
- Provide each child with pattern blocks.
- Tell the children to match the pattern blocks to the outlines on the puzzles.
- At this level, it is more common to see children intentionally flip and rotate shapes before placing them on the puzzle.
- It is important to allow children to use trial and error to complete the puzzle, however, also monitor if a child is becoming frustrated or completes the puzzle easily. Consider moving these children to a more appropriate level when choosing their next activity.
- As a child finishes a puzzle, give him or her a more difficult puzzle.

Note on Using Pattern Blocks to Teach Spatial Visualization: While these activities were developed for Shape Composition, if using them for Spatial Visualization remember to emphasize slides, flips, and turns. These are an example of ways to use pattern blocks to teach Spatial Visualization, but if your children are on a higher or lower level of Shape Composition, that's OK! They can work on Spatial Visualization at their level.

Materials Needed

- Pattern Block Puzzles (Shape Composer - Set 1)
- Pattern Blocks
- Additional pattern block puzzles are available on the internet and in books

---

Printable Activity

- Spanish PDF
- English PDF

Figure 10.3e (Continued).

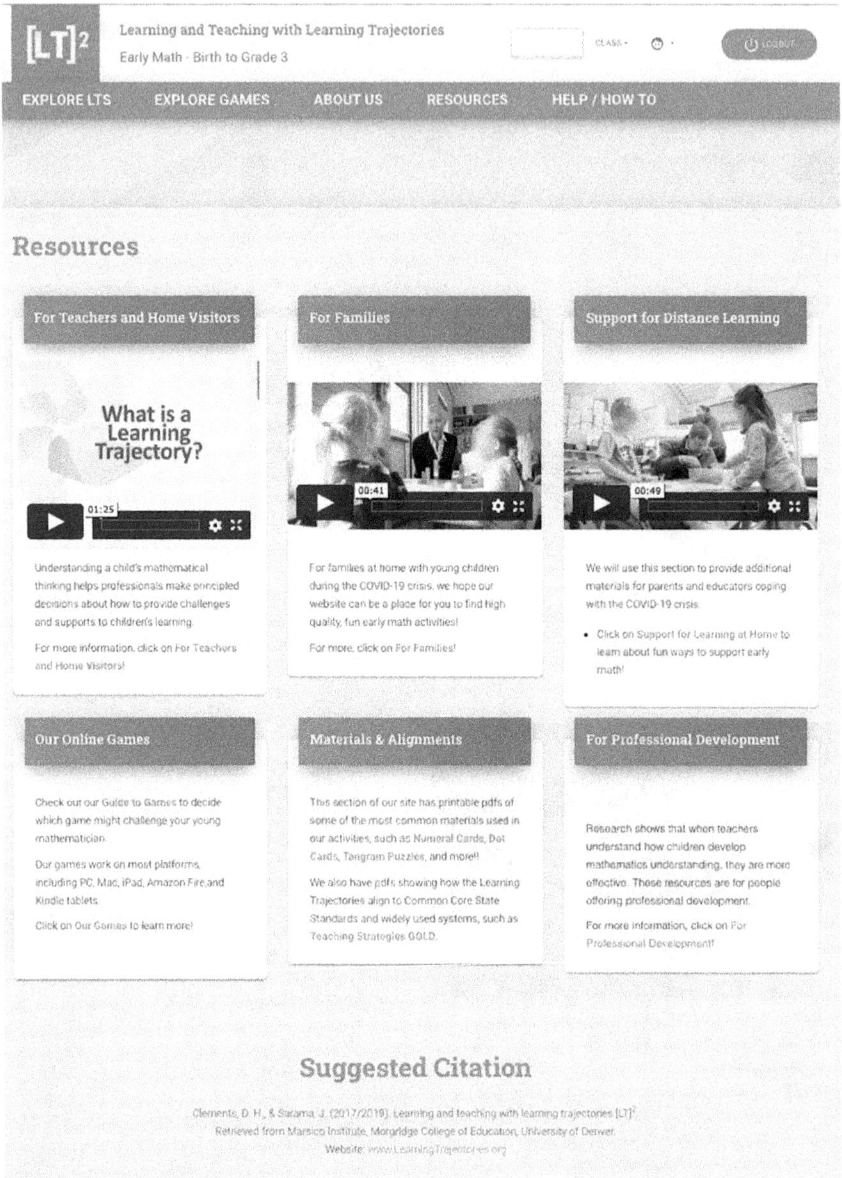

Figure 10.3f (Continued).

Evaluation of the PD approach

In a series of studies, we evaluated the effects of the PD model. As stated previously, the first two were quite successful but were smaller in scope and were formative evaluations of the model (Clements & Sarama, 2007, 2008). In the large-scale study, whose PD implementation was described previously, we evaluated the effects of scaling up the model on teachers' practices in early mathematics in the short and long term and whether teachers' practices mediated the effects of the intervention on students' outcomes.

Methods

Original adult participants were 106 pre-school teachers from 42 schools. Schools were randomly assigned, by site, to one of three treatment groups: two were experimental (72 teachers in 26 schools), with no differences relevant to this study (differences in the experimental treatments began after the pre-school year), and one was the business-as-usual control (34 teachers in 16 schools). Schools and teachers were not volunteers—a common limitation in many such studies (Clements et al., 2011). Schools served low-resource, ethnically diverse children. Their administration required every teacher to participate. Teachers in the treatment group were required to attend all PD sessions (control teachers received PD after the end of the study). Multiple measures and statistical procedures were used, including one-on-one assessment of children, questionnaires, and direct observations of teachers and classrooms with two instruments (psychometric pluralism, Blömeke et al., 2015).

Results

The implementation increased children's mathematical competencies (effect size, $d = .72$ standard deviations, Clements et al., 2011) as well as showing transfer to other domains, such as oral language (Sarama, Lange, Clements, & Wolfe, 2012). Here, we emphasize the impact on teachers.

Quantitative analyses showed that teachers in experimental groups showed significantly higher quality practices as measured by observational instruments and their practices mediated the effect of treatment on children's math outcome scores. Practices that were found to be particularly relevant in analyses included the "classroom culture" score—the quality of teacher/child interactions, as well as teacher attributes such as confidence, joy, curiosity, and enthusiasm.

To check the sustainability of these benefits, we visited the classrooms two years after the project ended. Although a logical expectation would be that, after the cessation of external support and PD provided by the intervention, teachers would show a pattern of decreasing levels of fidelity, these teachers actually demonstrated increasing levels of fidelity. They sustained high levels of fidelity of implementation two years past exposure to curriculum training

(Clements, Sarama, Wolfe, & Spitler, 2015). The main variable that predicted high fidelity was teachers noticing and appreciating children's learning of substantial mathematics.

The pre-school teachers in this study were experienced and well educated; yet, when asked to describe self-changes after participating in learning trajectories-based PD, their most salient and dominant response was surprise at their children's mathematical capabilities. Two-thirds of the teachers who were interviewed focused on their newfound understanding of pre-schoolers' mathematical capabilities or an increased expectation for pre-schoolers' mathematical achievement.

Many of these teachers explicitly described their changed understanding of children's ability to reason, explicate, justify, and problem solve; lead the way; and understand complex ideas, make connections, and understand relationships. When asked how they changed, teachers kept the focus on the children, explaining the change in their perceptions of children, and thus maintaining the child-centered focus of the approach. A related theme addressed changes in how they taught various groups (gender, ethnic/race, ability), always in the direction of equity and appreciation of all children's ability to learn mathematics. For example, some expressed surprise that certain groups of children (minority) were not only "ready to learn" but quite capable of advanced mathematical thinking.

In addition to the theme of children's mathematical capabilities, teachers described how they changed through learning about (a) learning trajectories, (b) how to teach mathematics to young children, (c) how important small group instruction was to implementing LT instruction, (d) how to use formative assessment in early mathematics, (e) how to augment small group instruction, and (f) how early mathematics was suitable for young children when incorporating mathematical activities through play.

More remarkable, we returned *six years* after the external intervention, three times the duration of the original sustainability analyses (Sarama et al., 2016). At this point, data were available only for 28 teachers (who were not significantly different from the others in demographics, years of experience, or any other variable we checked). The pattern of sustained fidelity to the curriculum was sustained with only one exception. Computer usage declined, probably because the initial licenses for the computer program lapsed. Results on other variables support the original sustainability findings. For example, teachers scored almost identically as in earlier time points on featuring mathematics materials in their classrooms, using everyday math activities, extending and enhancing the written activities, and teaching those activities with fidelity. They encouraged students to actively think, reason, solve problems, or reflect and involved and supported students in discussions of mathematics about the same or even more than at earlier time points. They maintained or increased their ability to use formative assessment (Sarama et al., 2016). Again, teachers taught the curriculum with *increasing* fidelity internalizing the learning trajectories approach (cf. Timperley, Wilson, Barrar, & Fung, 2007).

Conclusions

The learning trajectories' PD approached appeared to provide teachers with a coherent program of teaching and learning, which may have promoted the high-quality practice found in this study (Wilson et al., 2013). That is, teachers developed pedagogical *competence* based on specific *competencies* (Blömeke et al., 2015) from the learning trajectories training.

Thus, the first implication is that such a coherent model may provide the conditions for promoting high-quality instruction. The second is that such instruction is sustained for years (Clements & Sarama, 2014). The third is that the model may be particularly beneficial in addressing a climate of low expectations (Johnson & Fargo, 2010), as teachers increase their understanding of the capacities of all students to learn mathematics. Teachers who demonstrate sustained fidelity of implementation to a program that has demonstrated improved student achievement for each student will have a positive impact on many more students than teachers who implement with fidelity only during treatment. Thus, a fourth implication is that helping teachers develop the skills and practices necessary to perceive and document their students' learning may be an effective way to maintain and even increase the fidelity of implementation. These positive perceptions of learning may be especially important in motivating teachers to productively face the challenges inherent in fully implementing all aspects of the curriculum. Success breeds success, and such changes in practice may lead to positive changes in beliefs (e.g., Showers et al., 1987), again, resulting in mutually reinforcing changes in beliefs and practices (Caudle & Moran, 2012) rather than conflicts between them (cf. Einarsdottir, 2003), supporting bidirectional path between teachers' knowledge of practices and their beliefs.

These results confirm that effectiveness depends on the development of teachers' skills and knowledge. A total of 50–70 hours of PD—we provided about 75 hours—is consistent with previous research documenting what is necessary to achieve measurable effectiveness (Yoon, Duncan, Lee, Scarloss, & Shapley, 2007). Situating the materials not just in the classroom but also each school is important for related but additional reasons. Because it establishes a cultural practice and provides peer support, school-based implementation supports both fidelity and sustainability of a new curriculum (Zaslow, Tout, Halle, Vick, & Lavelle, 2010) and reaches over 90% of students in the school (not just one classroom, hoping for "spillover").

Learning trajectories can play a substantial positive role in both teachers' and children's learning. Experience with the learning trajectories generated fundamental change in teachers, as evidenced by their responses (100% reported change) and by observation. They were more aware and appreciative of young children's mathematical capabilities and of the pedagogical use of all three components of learning trajectories: mathematical goals and content, developmental progressions, and instructional tasks and strategies (cf. Jacobs & Empson, 2016).

Acknowledgment

This research was supported by the Institute of Education Sciences, US Department of Education, through grants R305K05157 and R305A110188; the National Science Foundation, through grants ESI-9730804 and REC-0228440; the Heising-Simons Foundation through Grants #2013-79 and #2015-157; and the Bill & Melinda Gates Foundation through Grant #OPP1118932. The opinions expressed are those of the authors and do not represent the views of the funders. The authors wish to express appreciation to the school districts, teachers, and students who participated in this research.

Note

1 This term is used in some countries; in others, the term "curriculum" is used to indicate such educational objectives.

References

Association of Mathematics Teacher Educators. (2017). *AMTE standards for mathematics teacher preparation*. Raleigh, NC: AMTE.

Ball, D. L., & Cohen, D. K. (1999). *Instruction, capacity, and improvement*. Philadelphia, PA: Consortium for Policy Research in Education, University of Pennsylvania.

Baroody, A. J., & Ginsburg, H. P. (1990). Children's mathematics learning: A cognitive view. In R. B. Davis, C. A. Maher, & N. Noddings (Eds.), *Constructivist views on the teaching and learning of mathematics. Journal for Research in Mathematics Education Monograph Number 4* (pp. 79–90). Reston, VA: National Council of Teaching of Mathematics.

Battista, M. T. (2013, April). Elementary teachers' learning, understanding, and classroom use of learning progressions. In *Paper presented at the National Council of Teachers of Mathematics*, Denver, CO.

Björklund, C., & Barendregt, W. (2016). Teachers' pedagogical mathematical awareness in diverse child-age-groups. *Nordic Studies in Mathematics Education, 21*(4), 115–133. https://doi.org/10.1080/00313831.2015.1066426

Blömeke, S., Gustafsson, J.-E., & Shavelson, R. J. (2015). Beyond dichotomies: Competence viewed as a continuum. *Zeitschrift für Psychologie, 223*(1), 3–13. https://doi.org/10.1027/2151-2604/a000194

Bobis, J., Clarke, B. A., Clarke, D. M., Gill, T., Wright, R. J., Young-Loveridge, J. M., & Gould, P. (2005). Supporting teachers in the development of young children's mathematical thinking: Three large scale cases. *Mathematics Education Research Journal, 16*(3), 27–57. https://doi.org/10.1007/BF03217400

Brendefur, J. L., Strother, S., Thiede, K., Lane, C., & Surges-Prokop, M. (2013). A professional development program to improve math skills among preschool children in head start. *Early Childhood Education Journal, 41*(3), 187–195. https://doi.org/10.1007/s10643-012-0543-8

Bryk, A. S., Sebring, P. B., Allensworth, E., Suppescu, S., & Easton, J. Q. (2010). *Organizing schools for improvement: Lessons from Chicago*. Chicago, IL: University of Chicago Press.

Carpenter, T. P., Fennema, E. H., Franke, M. L., Levi, L., & Empson, S. B. (2014). *Children's mathematics: Cognitively guided instruction* (2nd ed.). Portsmouth, NH: Heinemann.

Carpenter, T. P., Franke, M. L., Jacobs, V. R., Fennema, E. H., & Empson, S. B. (1998). A longitudinal study of invention and understanding in children's multidigit addition and subtraction. *Journal for Research in Mathematics Education, 29*(1), 3–20. https://doi.org/10.2307/749715

Caudle, L., & Moran, M. J. (2012). Changes in understandings of three teachers' beliefs and practice across time: Moving from teacher preparation to in-service teaching. *Journal of Early Childhood Teacher Education, 33*, 38–53.

Clarke, B. A. (2008). A framework of growth points as a powerful teacher development tool. In D. Tirosh & T. Wood (Eds.), *Tools and processes in mathematics teacher education* (pp. 235–256). Rotterdam: Sense.

Clarke, D. M., Cheeseman, J., Clarke, B., Gervasoni, A., Gronn, D., Horne, M., McDonough, A., Montgomery, P. Rowley, G., & Sullivan, P. (2001). Understanding, assessing and developing young children's mathematical thinking: Research as a powerful tool for professional growth. In J. Bobis, B. Perry & M. Mitchelmore (Eds.), Numeracy and beyond (Proceedings of the 24th Annual Conference of the Mathematics Education Research Group of Australasia, Vol. 1) (pp. 9–26). Reston, Australia: MERGA.

Clements, D. H., & Sarama, J. (2004). Learning trajectories in mathematics education. *Mathematical Thinking and Learning, 6*, 81–89. https://doi.org/10.1207/s15327833mtl0602_1

Clements, D. H., & Sarama, J. (2007). Effects of a preschool mathematics curriculum: Summative research on the *Building Blocks* project. *Journal for Research in Mathematics Education, 38*(2), 136–163. https://doi.org/10.2307/748360

Clements, D. H., & Sarama, J. (2008). Experimental evaluation of the effects of a research-based preschool mathematics curriculum. *American Educational Research Journal, 45*(2), 443–494. https://doi.org/10.3102/0002831207312908

Clements, D. H., & Sarama, J. (2014). Learning trajectories: Foundations for effective, research-based education. In A. P. Maloney, J. Confrey, & K. H. Nguyen (Eds.), *Learning over time: Learning trajectories in mathematics education* (pp. 1–30). New York, NY: Information Age Publishing.

Clements, D. H., Sarama, J., Spitler, M. E., Lange, A. A., & Wolfe, C. B. (2011). Mathematics learned by young children in an intervention based on learning trajectories: A large-scale cluster randomized trial. *Journal for Research in Mathematics Education, 42*(2), 127–166. https://doi.org/10.5951/jresematheduc.42.2.0127

Clements, D. H., Sarama, J., Wolfe, C. B., & Spitler, M. E. (2015). Sustainability of a scale-up intervention in early mathematics: Longitudinal evaluation of implementation fidelity. *Early Education and Development, 26*(3), 427–449. https://doi.org/10.1080/10409289.2015.968242

Confrey, J., Maloney, A. P., Nguyen, K. H., & Rupp, A. A. (2014). Equipartitioning: A foundation for rational number reasoning. Elucidation of a learning trajectory. In A. P. Maloney, J. Confrey, & K. H. Nguyen (Eds.), *Learning over time: Learning trajectories in mathematics education* (pp. 61–96). New York, NY: Information Age Publishing.

Einarsdottir, J. (2003). Beliefs of early childhood teachers. In O. N. Saracho & B. Spodek (Eds.), *Studying teachers in early childhood settings* (pp. 113–134). Greenwich, CT: Information Age Publishing.

Elmore, R. F. (1996). Getting to scale with good educational practices. *Harvard Educational Review, 66*, 1–25.

Feltovich, P. J., Spiro, R. J., & Coulson, R. L. (1997). Issues of expert flexibility in contexts characterized by complexity and change. In P. J. Feltovich, K. M. Ford, & R. R. Hoffman (Eds.), *Expertise in context: Human and machine* (pp. 125–146). Cambridge, MA: The MIT Press.

Franke, M. L., Carpenter, T. P., Levi, L., & Fennema, E. H. (2001). Capturing teachers' generative change: A follow-up study of professional development in mathematics. *American Educational Research Journal, 38*, 653–689.

Gerofsky, P. R. (2015). Why Asian preschool children mathematically outperform preschool children from other countries. *Western Undergraduate Psychology Journal, 3*(1). Retrieved May 31, 2015 from http://ir.lib.uwo.ca/wupj/vol3/iss1/11

Guskey, T. R. (Ed.) (2000). *Evaluating professional development.* Thousand Oaks, CA: Corwin Press.

Hall, G. E., & Hord, S. M. (2001). *Implementing change: Patterns, principles, and potholes.* Boston, MA: Allyn and Bacon.

Heck, D. J., Weiss, I. R., Boyd, S., & Howard, M. (2002). Lessons learned about planning and implementing statewide systemic initiatives in mathematics and science education. In *Paper presented at the American Educational Research Association*, New Orleans, LA. www.horizon-research.com/public.htm

Institute of Medicine (IOM) and National Research Council (NRC) (2015). *Transforming the workforce for children birth through age 8: A unifying foundation.* Washington, DC: National Academy Press.

Jacobs, V. R., & Empson, S. B. (2016). Responding to children's mathematical thinking in the moment: An emerging framework of teaching moves. *ZDM Mathematics Education, 48*, 185–197. https://doi.org/10.1007/s11858-015-0717-0

Johnson, C. C., & Fargo, J. D. (2010). Urban school reform enabled by transformative professional development: Impact on teacher change and student learning of science. *Urban Education, 45*(1), 4–29.

Klingner, J. K., Ahwee, S., Pilonieta, P., & Menendez, R. (2003). Barriers and facilitators in scaling up research-based practices. *Exceptional Children, 69*, 411–429.

MacDonald, A., Davies, N., Dockett, S., & Perry, B. (2012). Early childhood mathematics education. In B. Perry, T. Lowrie, T. Logan, A. MacDonald, & J. Greenlees (Eds.), *Research in mathematics education in Australasia: 2008–2011* (pp. 169–192). Rotterdam, The Netherlands: Sense Publishers.

McCoy, D. C., Salhi, C., Yoshikawa, H., Black, M., Britto, P., & Fink, G. (2018). Home- and center-based learning opportunities for preschoolers in low- and middle-income countries. *Children and Youth Services Review, 88*, 44–56. https://doi.org/10.1016/j.childyouth.2018.02.021

Mullis, I. V. S., Martin, M. O., Foy, P., & Arora, A. (2012). *TIMSS 2011 international results in mathematics.* Chestnut Hill, MA: TIMSS & PIRLS International Study Center, Boston College.

National Research Council. (2009). *Mathematics learning in early childhood: Paths toward excellence and equity.* Washington, DC: National Academy Press.

NCTM. (2006). *Curriculum focal points for prekindergarten through grade 8 mathematics: A quest for coherence.* Reston, VA: National Council of Teachers of Mathematics.

NGA/CCSSO. (2010). *Common core state standards.* Washington, DC: National Governors Association Center for Best Practices, Council of Chief State School Officers.

OECD. (2014). *Strong performers and successful reformers in education – Lessons from PISA 2012 for the United States.* Paris, France: OECD Publishing.

Pellegrino, J. W. (2007). From early reading to high school mathematics: Matching case studies of four educational innovations against principles for effective scale up. In B. Schneider & S.-K. McDonald (Eds.), *Scale up in practice* (pp. 131–139). Lanham, MD: Rowman and Littlefield.

Perry, B. (2010). Mathematical thinking of preschool children in rural and regional Australia: An overview. *Journal of Australian Research in Early Childhood Education, 16*(2), 1–12.

Piasta, S. B., Logan, J. A. R., Pelatti, C. Y., Capps, J. L., & Petrill, S. A. (2015). Professional development for early childhood educators: Efforts to improve math and science learning opportunities in early childhood classrooms. *Journal of Educational Psychology, 107*(2), 407–422. https://doi.org/10.1037/a0037621

Polly, D., McGee, J., Wang, C., Martin, C., Lambert, R., & Pugalee, D. K. (2015). Linking professional development, teacher outcomes, and student achievement: The case of a learner-centered mathematics program for elementary school teachers. *International Journal of Educational Research, 72*, 26–37. https://doi.org/10.1016/j.ijer.2015.04.002

Sarama, J. (2002). Listening to teachers: Planning for professional development. *Teaching Children Mathematics, 9*(1), 36–39.

Sarama, J., & Clements, D. H. (2009). *Early childhood mathematics education research: Learning trajectories for young children.* New York, NY: Routledge.

Sarama, J., & Clements, D. H. (2019). The building blocks and TRIAD projects. In P. Sztajn & P. H. Wilson (Eds.), *Learning trajectories for teachers: Designing effective professional development for math instruction* (pp. 104–131). New York, NY: Teachers College Press.

Sarama, J., Clements, D. H., Starkey, P., Klein, A., & Wakeley, A. (2008). Scaling up the implementation of a pre-kindergarten mathematics curriculum: Teaching for understanding with trajectories and technologies. *Journal of Research on Educational Effectiveness, 1*(1), 89–119. https://doi.org/10.1080/19345740801941332

Sarama, J., Clements, D. H., Wolfe, C. B., & Spitler, M. E. (2016). Professional development in early mathematics: Effects of an intervention based on learning trajectories on teachers' practices. *Nordic Studies in Mathematics Education, 21*(4), 29–55.

Sarama, J., Lange, A., Clements, D. H., & Wolfe, C. B. (2012). The impacts of an early mathematics curriculum on emerging literacy and language. *Early Childhood Research Quarterly, 27*(3), 489–502. https://doi.org/10.1016/j.ecresq.2011.12.002

Showers, B., Joyce, B., & Bennett, B. (1987). Synthesis of research on staff development: A framework for future study and a state-of-the-art analysis. *Educational Leadership, 45*(3), 77–87.

Shulman, L. S. (1986). Those who understand: Knowledge growth in teaching. *Educational Researcher, 15*(2), 4–14.

Stosich, E. L. (2016). Joint inquiry: Teachers' collective learning about the Common Core in high-poverty urban schools. *American Educational Research Journal, 53*(6), 1698–1731. https://doi.org/10.3102/0002831216675403

Sztajn, P., Confrey, J., Wilson, P. H., & Edgington, C. (2012). Learning trajectory based instruction: Toward a theory of teaching. *Educational Researcher, 41*, 147–156. doi: 10.3102/0013189X12442801

Timperley, H., Wilson, A., Barrar, H., & Fung, I. (2007). *Teacher professional development: Best Evidence Synthesis iteration.* Wellington, New Zealand: Ministry of Education.

Weiss, I. R. (2002). Systemic reform in mathematics education: What have we learned? In *Paper presented at the research presession of the 80th annual meeting of the National Council of Teachers of Mathematics*, Las Vegas, NV.

Wilson, P. H., Mojica, G. F., & Confrey, J. (2013). Learning trajectories in teacher education: Supporting teachers' understandings of students' mathematical thinking. *The Journal of Mathematical Behavior, 32*(2), 103–121. https://doi.org/10.1016/j.jmathb.2012.12.003

Yoon, K. S., Duncan, T., Lee, S. W.-Y., Scarloss, B., & Shapley, K. L. (2007). *Reviewing the evidence on how teacher professional development affects student achievement (Issues & Answers Report, REL 2007–No. 033).* Retrieved January 11, 2015 from http://ies.ed.gov/ncee/edlabs

Zaslow, M. J., Tout, K., Halle, T. G., Vick, J., & Lavelle, B. (2010). *Towards the identification of features of effective professional development for early childhood educators: A review of the literature.* Retrieved November 1, 2014 from http://www2.ed.gov/rschstat/eval/professional-development/literature-review.pdf

Chapter 11

Pre-school teachers' ways of promoting mathematical learning in picture book reading

Camilla Björklund, Hanna Palmér

Introduction

Picture books are commonly used with pedagogical intentions in pre-school education. Several studies support picture books' usefulness as pedagogical tools, as they relate to children's everyday experiences and thereby may foster children's mathematical knowledge (Hassinger-Das, Jordan, & Dyson, 2015; Jennings, Jennings, Richey, & Dixon-Krauss, 1992). Nevertheless, decoding the meaning of, or the notions involved in, a narrative is a complex undertaking (Shiyan, Björklund, & Pramling Samuelsson, 2018) and it cannot be taken for granted that mathematical meaning is mediated even if it is highlighted in both text and pictures in a picture book.

In this chapter, we address the issue of reading picture books for pedagogical purposes. In the study referred to, a picture book was designed for the purpose of raising mathematical attention. In this study, picture book reading is considered to be a situation for the potential teaching and learning of mathematics, involving the teacher, the child, and the mathematical content in the book. In this triad, the teacher can facilitate learning by directing the child's attention towards mathematical aspects in the book. But, what do teachers need to know in order to teach mathematics in book reading? This comes down to the essential question of teacher competence, which is a challenging endeavour that has to be seen in the light of the messy complex situations that early childhood education often constitutes (Blömeke, Gustafsson, & Shavelson, 2015). In this chapter, we analyse video documentations of three pre-school teachers reading the specially designed picture book to 13 pre-schoolers (3–5 years old). These book-reading sessions are analysed with a focus on teachers' mathematical knowledge *for* teaching (MKT) (Ball, Thames, & Phelps, 2008). More specifically, we direct attention to the questions how teachers' MKT is made visible in book reading with pre-schoolers, and how teachers' MKT may be related to children's learning opportunities. Our overarching aim is thereby to contribute a discussion on the relation between teacher knowledge and potential learning outcomes.

DOI: 10.4324/9781003172529-11

Swedish pre-school

Swedish pre-school has a long tradition of social pedagogy (Bennett & Tayler, 2006), whereby care, socialisation, and learning are considered a coherent whole. In recent revisions of the Swedish national curriculum (National Agency for Education, 2018), however, it is made explicit that teaching is to take place in pre-school. Even though the curriculum encompasses several goals to strive for, including mathematical ones, there are no standards for children to attain.

The attendance rate in Swedish pre-school is 95% among 4–5-year olds (National Agency for Education, 2019). It is common to find children of different ages in the same pre-school groups. This is a demanding setting for teaching, as there is a great variance in skills and knowledge among the children. Thus, the teachers need to be sensitive to and attend individual children's competences. Furthermore, in Swedish pre-school practice, it is traditionally important that teachers attend to children's initiatives, which demands high levels of flexibility in order to adhere to these initiatives while simultaneously maintaining focus on the goals of the teaching act.

Picture books in mathematics education

Illustrated stories have shown to increase the interaction between child and reader during joint reading and also the number of story events remembered by the children after their book reading (Greenhoot, Beyer, & Curtis, 2014). In picture books, the pictures and narrative usually interact as the pictures provide support to the narrative. However, illustrations are not necessarily beneficial per se and should thus be carefully designed to avoid distraction and drawing attention to irrelevant features (Ward, Mazzocco, Bock, & Prokes, 2017). Also, if the narratives include many different features, these can draw the children's attention away from the mathematical content (Pramling & Pramling Samuelsson, 2008). To facilitate mathematical learning, according to Van den Heuvel-Panhuizen and Elia (2012), picture books need a good story that engages children, the mathematics ought to be available but not too blatant, and a wide range of mathematical topics are preferred to overly narrow mathematical content.

At the same time as there seems to be pedagogical potential in using picture books, the research is ambiguous regarding the teacher's role. Studies have shown that picture books of high literary quality, albeit not written for the purpose of teaching mathematics, promote mathematical thinking among young children even without prompting by the reader (Van den Heuvel-Panhuizen & Van den Boogaard, 2008). However, a recent study (Björklund & Palmér, 2020) found significant evidence for mathematical reasoning of higher quality in dialogic reading sessions with mutual initiatives in extending content meaning. Thus, there are contradictory findings in the research concerning the extent to which the teacher/reader influences the possible learning outcome. We thereby argue that there is more to know about the teacher's role in picture book reading, going beyond a simple dichotomy of teacher interaction or not, and instead attending to *how* the teacher acts and based on *what* MKT s/he possesses.

Research on mathematics teachers' knowledge

There are several studies on the knowledge teachers have and use and what knowledge that is needed to teach mathematics. According to Sowder (2007), there is a difference between knowing mathematics and being able to teach mathematics. In order to teach mathematics, teachers need to have knowledge not only about mathematics but also mathematics teaching. According to Davis and Simmt (2006), *mathematics-for-teaching* is not a matter of more, or a greater depth than, ordinary mathematics; it is qualitatively different. However, there is no consensus in the research regarding what mathematics teachers actually need to know. This may be because knowledge is a relative notion, dependent on the context within which it operates (Wilson & Cooney, 2002). Similarly, Llinares and Krainer (2006) write that research has shown how complex the question of teacher knowledge is, which is why it cannot be treated as independent but rather as situated in the context of teaching.

Mathematical knowledge for teaching

MKT is a framework that refers to the special kind of knowledge needed for teaching mathematics. The framework contains of six parts: common content knowledge, specialised content knowledge, knowledge at the mathematical horizon, knowledge of content and students, knowledge of content and teaching, and knowledge of content and curriculum (Ball et al., 2008).

Common content knowledge refers to mathematical knowledge that is used in teaching in a way similar to how it is used in settings other than teaching. Pre-school teachers use mathematics in many situations to organise their group of pre-schoolers, make a schedule for the daily plans, and so on. But this kind of knowledge is not enough for teaching young children mathematics, even though it may serve as an example of when mathematical knowledge is used. Specialised content knowledge, on the other hand, refers to mathematical knowledge that is unique to teaching. This knowledge is qualitatively different and means more than knowing mathematics in everyday life or as an academic discipline. For instance, specialised content knowledge includes the knowledge needed to analyse wrong answers, develop alternative explanations, and choose usable definitions. Knowledge at the mathematical horizon refers to knowledge of how mathematics is related to other topics included in the curriculum, for example knowledge of how symbols are used in other settings than mathematical ones.

Knowledge of content and students combines knowledge of students with that of mathematics, for example how students typically use and understand different representations or knowledge about mathematical operations that often pose difficulties for them. This means that a teacher can anticipate different answers and prepare for how to facilitate children's learning, which naturally differs among children of different ages and with different experiences. Particularly, this is at the core of pre-school teachers' professional competence as they work with a group of children that is heterogeneous regarding both age and experiences (see the distinct feature of Swedish pre-school lacking standards and including mixed age groups).

Knowledge of content and teaching combines knowledge of teaching with knowledge of mathematics, for example how to explain and which examples to use, and how and when to use multiple representations. This kind of knowledge is challenging in a pre-school setting, as (at least in the Swedish context) the teaching may be conducted for example outdoors, through play and games, and through spontaneous activities. This dynamic nature of mathematics education requires the teacher to have a broad and flexible repertoire of examples, representations, and ways of interacting with young children. Knowledge of content and curriculum is furthermore knowledge about classroom organisation, teaching environments, and both one's own curriculum as well as those of other grades and subjects.

The MKT framework applies to mathematics teachers at all school levels but becomes particularly demanding, we suggest, in early childhood education. Similar suggestions have been made by Mosvold, Bjuland, Fauskanger, and Jakobsen (2011), as young children's knowledge is at an emerging stage and cannot be interpreted through graphical expressions such as formal mathematical symbols. Instead, their knowledge comes through in actions and subtle expressions of reasoning, often in non-verbal manners.

Variation theory of learning

The MKT framework was empirically developed to investigate the nature of professionally oriented subject-matter knowledge of mathematics. However, according to Ball et al. (2008), it is difficult to identify what mathematical knowledge is specific to teachers. To overcome these difficulties and further deepen our understanding of how MKT appears in the teaching of young children, we add the variation theory of learning (VT) (Marton, 2015) to the framework for our study.

According to VT, learning always entails the learning of something, and it is how this something is discerned, or made possible to discern, that is interesting to explore in studies influenced by VT. Thus, VT attends to the conditions for learning that appear to be necessary when a child encounters a learning object. How a learning object and aspects of that learning object are possible to discern by a child constitutes a *potential learning outcome*. Given that a teacher has different kinds of MKT, VT adds *how* the teacher's enactment will make it possible for the child to discern aspects of a learning object that s/he has not experienced before.

VT provides a framework for how a teacher can enable a child to discern desired aspects. Any learning object can be seen as constituting certain aspects.

Through different *patterns of variation* and *invariance*, it is possible for the child to experience the learning object in more advanced ways. For example, in the case of numbers, there are some aspects of numbers that the child has to become aware of, such as the cardinality of numbers, ordinality, and that numbers may have a specific relation to other numbers in terms of part-whole relations. If a child has only discerned numbers as words in a sequence, s/he cannot make use of numbers for problem solving. Numbers are then perhaps best used when playing hide-and-seek as a way to keep the time for the players to find a hiding place. But to figure out how many players are still missing, the child also needs to be aware of the cardinal aspect of numbers and number relations. This knowledge of how numbers are constituted as a complex construct of different aspects that each bear significant meaning can be used in designing teaching in which critical aspects are made discernible by contrasting their appearing features (such as contrasting sets of two with sets of three while the features of the objects are kept invariant). This calls for knowledge of what aspects the child needs to discern in order to "see" numbers in a more advanced way (specialised content knowledge) and knowledge of how to make this possible in teaching (knowledge of content and teaching) for a specific child (knowledge of content and students).

Method

To answer our research question, we designed a study in which pre-school teachers read a specially designed picture book to their pre-schoolers. Video observations of these reading sessions are the main data for analysis. However, the picture book is also an essential component of the study as it constitutes the mathematical content to which the teacher-child interaction is directed.

Designing the picture book

To create the best conditions for reading picture books for pedagogical purposes, a picture book adapted to the aim of the study was developed by the authors of this chapter. The book was designed to adhere to the evaluation criteria of Van den Heuvel-Panhuizen and Elia (2012): the mathematical content is correct and presented in a meaningful context; connections within mathematics and to children's interests are possible; understanding is possible at different levels and future concept development anticipated; and opportunities for cognitive, emotional, or physical engagement are offered. Further, when designing the picture book, we used VT (Marton, 2015), so that significant aspects (known from earlier research and literature in the field) would be possible to discern through patterns of variation incorporated in and between the pictures.

The picture book is about a cat living in a house in the forest. One day, the cat is hungry and goes out to find some food. During the search for food, the cat meets one friend on each new spread. Each animal has a different symbol on its sweater. For every new friend, berries visualised in different collections are found. A reoccurring feature on all spreads is collections of objects that are designed in different patterns to attract attention to sets of items and the part-whole relation of numbers, such as different berries composed as sets of coloured circles.

Participating pre-schools

Pre-school teachers in a network for early mathematics learning and teaching in Sweden were invited to participate in the study. The teachers were given a brief introduction to contemporary research on picture books in early mathematics education, and more detailed instructions on how to conduct the empirical study upon volunteering to participate. Three teachers volunteered. The participating children's legal representatives were given written information about the study and approved their children's participation in line with the ethical guidelines provided by the Swedish Research Council (2017). The children were aged 3–5 years and were all fluent Swedish speakers.

Pre-school teacher "Polly" had 8 years of teaching experience and conducted the individual reading with six children. Pre-school teacher "Yulia" had 15 years of teaching experience and conducted the individual reading with five children. Pre-school teacher "Annie" had 34 years of teaching experience and conducted the individual reading with two children.

The book-reading sessions

To create the best conditions for the teachers to listen to the children's responses and to incorporate them in the conversations, the book was read to one child at a time. The teachers were not given any scripts for these book-reading sessions but were instead instructed to read in a "common pedagogical fashion". This approach, previously used by for instance Van den Heuvel-Panhuizen and Elia (2013), is intended to encourage the reader to ask open-ended questions, to avoid yes/no or overly pointed questions, and to respond, expand, and initiate discussions with the child without asking too many questions. This is much in line with the "common pedagogical fashion" in Swedish pre-school.

Analysis

To contribute to the discussion about the relation between teacher knowledge and potential learning outcomes, we analysed authentic book-reading sessions how teachers' MKT was made visible, and how teachers' MKT may be related to children's learning opportunities. We focused especially on *knowledge of content and students* and *knowledge of content and teaching*. Similar to Ball and colleagues (2008), we focused on the nature of professionally oriented subject-matter knowledge in mathematics by studying actual mathematics teaching and the mathematical problems that arise in teaching.

The book-reading sessions were video-recorded by the pre-school teachers. We implemented the *critical incident technique* (Butterfield, Borgen, Amundson, & Maglio, 2005) to distinguish episodes involving mathematics-related utterances or gestures. We then conducted a thorough qualitative analysis of these critical incidents, focusing on how different ways the pre-school teachers acted were related to the children's initiatives and actions, as expressions of aspects that had been brought to the forefront and thus made possible to explore. From this discussion, we conclude what mathematical knowledge is demanded by the work teachers do. The episodes from the empirical data in the following section are chosen as examples to illustrate the larger sample of observations.

Results

In the following, we present examples from the observed reading sessions and discuss how the teachers act and how their expressed MKT informs what is made possible for the children to learn.

As mentioned earlier, the Swedish pre-school context is influenced by a high degree of freedom for both teachers and children to take initiatives. The following example highlights the sensitivity to the child's interest and competence and the book reading, as well as the need for teachers' flexibility and ability to uphold the content in focus in the teaching:

Polly (teacher) reads spread 4

ERIN: Oh, an eight! I can shape myself as an eight.
POLLY: Can you? Let me see.
　　Erin gets up and walks across the floor: Mm, one turn like that, and mm…
POLLY: That's exactly like an eight.
ERIN: Now I'm gonna try doing that one (points at the triangle on the rabbit's sweater)
POLLY: Yes, exactly, what's it called?
　　Erin walks across the floor.
POLLY: Mm, out in the corners.
ERIN: And like this, and this (looks in the book and turns)
POLLY: Yes, and then you're back where you started, three sections. Just like that, one, two, three, what do we call it then?
ERIN: Three-edging [trekangel, a non-standard Swedish term].
POLLY: A triangle, that's how I say it, do you know that?
ERIN: Yes.
　　Polly reads the text on the spread.
ERIN: NOW I can do the counting (starts counting the berries in the picture).

In this episode, the child is literally moving away from the book reading to the geometrical shapes she found in the book's pictures. The book's narrative

Spread 4: "When the cat and the dog had walked for a while they met their friend the rabbit. The cat and the dog and the rabbit came to a clearing with cloudberries. – I love cloud berries, said the rabbit. The cat and the dog and the rabbit ate all the cloudberries. – I am still hungry, said the cat".

is directed mainly towards numbers, and most of the children (and teachers) involve themselves in various kinds of counting acts. This child breaks this pattern, challenging the direction of attention by focusing on geometrical shapes. The teacher encourages her imitation of the shapes through moving in similar shapes but at the same time maintains the connection to the picture in the book that inspired her. It can be assumed that this connection to the book is essential in order to return to the narrative and to continue reading and then direct the child's attention to other content. These kinds of episodes tell us that there are opportunities for mathematics learning in book reading, but it is a demanding task to uphold a child's attention to a learning object and in situations that emerge during the reading. Thus, there is a challenge involved in attending to children's initiatives and extending their experiences while simultaneously maintaining their attention to a learning object.

Many mathematical notions are novel to young children, and their full meaning is not yet discerned. To extend the meaning of certain notions the children have encountered and thus have some experience of, it is common for teachers to point out the notion in question so as to generalise and specify meaning.

2

Spread 1: "Once upon a time there was a cat living in a house in the forest".

Spread 2: "One day, the cat was hungry and went out to look for some food".

This is not as simple as it might seem, however, as it requires careful attention to the specific notion and on *how* to generalise. The following two episodes exemplify this distinction:

Polly (teacher) reads spread 2

POLLY: The cat lives in Forest road [Skogsvägen, in Swedish]. Can you see what number?

ERIK: Three. There's another three (pointing at the page number).

Polly (Turns the pages to the previous spread 1) Do you see any threes here?

ERIK: There (points at the clock).
POLLY: On the clock. Any more threes?
ERIK: There and there (points at the note on the wall and the calendar).
POLLY: There's a long number sequence, 1, 2, 3, 4.

First, the teacher encourages the child to direct his attention to a specific number and then to distinguish the same number symbol found on a different spread. The initial number has a specific meaning, being linked to a specific address, which the child can find as a sign on the house. Similarly, the number 3 has a specific – but different – meaning in the other examples. Thus, the teacher does not only ask the child to recognise number symbols but also to identify them in a context in which numbers make sense. The child recognises more of the same number, and the teacher expands their search to the previous spread where several number symbols can also be found. Here again, she situates the symbols that the child finds by naming their contexts, such as the clock on the wall. Furthermore, she not only includes the numbers' situated meaning but also widens their meaning as part of the number sequence found on the calendar. These acts may seem to be subtle but at a closer look are very important for the child's opportunities to learn about numbers. In order to make use of numbers, their different mathematical aspects have to be discerned, as do their context-bound meanings. Graphical symbols like numbers do not give away their mathematical or contextual meaning themselves but rather have to be explored and explained in the different contexts within which they appear. To be able to explore such abstract meanings, the child needs to be exposed to and thus allowed to discern these symbols and their different meanings. In the example above, this is made possible by the teacher's careful situating of the number symbols in the different contexts while also pointing out a mathematical context (the number sequence) as a variation of different contexts in which numbers appear and bear different meanings. While situating these different contexts the number (3) is kept invariant. While the specific differences in meaning are not explored in the brief example, the child is given opportunities (through the contrasting contexts) to identify the reoccurrence of numbers as a basis for later exploration. The choice to keep the three invariant is essential in this episode, and may be an expression of the teacher's knowledge of numbers' complexity in meaning, which directs her acts in ways that allow the child to discern certain features of numbers that are foundational for further exploration.

Taking up children's initiatives is a delicate act. At first glance, the teacher in the following episode seems to be considering the child's initiatives while situating his observations in a way that is reminiscent of the previous example. However, there is a qualitative (and, for learning opportunities, crucial) difference that may be drawn regarding teacher knowledge of connections within and between mathematical aspects.

6

Spread 3: "When she had walked for a while she met her friend the dog. The cat and the dog came to a clearing with blueberries. – I love blueberries, said the dog. The cat and the dog ate all blueberries. – I am still hungry, said the cat".

Yulia (teacher) reads spread 3
Tom points at each berry and counts to 20.

YULIA: What did you count?
TOM: 20.
YULIA: 20, was that how many blueberries there were? And lots of fir trees, tall and short ones.
TOM: One tall and lots of short. I'll count (points at each tree and counts to 20).

YULIA: Was there equally many?

TOM: 21 (points at the large tree).

YULIA: But do you recognize this? (points at the road sign).

TOM: You're allowed to drive 30.

YULIA: You're allowed to drive 30, good to have road signs.

This example is interesting in contrast to the previous one. The teacher indeed adheres to the child's initiatives, and the reading session is characterised by mutual interest and dialogue about content found in the book. Here, the content that could be developed into learning objects is initiated by both child and teacher; however, they remain single examples without connections. In sum, the child is acknowledged in what he identifies and already knows, but there is no extension of the meaning by connecting within and between mathematical notions. Successful teaching, on the other hand, involves the teacher's skill in extending the content meaning and thereby offering the child new aspects of a mathematical notion to be discerned. This requires the teacher to have specialised content knowledge; that is, to know the mathematics constituting the notion. In addition, the mathematics have to be within reach of discernment for the particular child (knowledge of content and student). In other words, the aspects provided to discern have to be connected to the child's previous understanding and need to be mathematically correct:

Annie (teacher) reads spread 3

HUGO: 1, 2, 3, 4, 5 (points at each blueberry in one group, repeats counting to 5 for each group).

ANNIE: Yes, five blueberries in each group. But did you wonder about what it says there? (points at the road sign). What do you think this looks like then?

HUGO: It looks like zero.

ANNIE: And this one? (points at the 3)

HUGO: Also a zero.

ANNIE: Okay, you think this also looks like a zero. Do you think they look the same?

HUGO: No, that's a two and that's a zero.

ANNIE: Well, you could think it's a two but actually it's the symbol for number three. And then there's a zero after, three and zero makes 30!

HUGO: 30.

In the above episode, the child is encouraged to read number symbols, which reveals that he has difficulties recognising them by their names (saying "a two" for the symbol 3). The teacher nevertheless continues describing the notion "30" as one symbol composed of two other symbols. It is thus questionable whether the child is given opportunities to extend his understanding of number symbols to two-digit numbers without first knowing the names of the first ten natural numbers. Explaining 30 as a combination of 3 and 0 without connecting it to the base-ten system is also an aspect that demands quite extensive pre-knowledge of numbers as units.

In the following episode, we observe one of the teachers in her acts of extending the opportunities to solve an arithmetic task by offering tools for structuring the mathematical content, in this case numbers' part-whole relations, to find missing units of a known whole. She does this without leaving the narrative of the book and thus maintaining the relevance of the arithmetical content. This is crucial in teaching in early childhood education, as young children who are generally given agency to orchestrate the direction of the learning content will simply not adhere to any irrelevant features of the narrative if they do not add some value to the story (see Pramling et al., 2019):

Polly (teacher) reads spread 8

ANDERS: I knew they were playing hide-and-seek because they're hiding!
POLLY: Did they hide? Where are they?
ANDERS: I wonder if the bear…
POLLY: Who do you see?

Anders points at the long tail behind the stone.

POLLY: Who was it that had such a tail?
ANDERS: I don't remember what it's called.

Spread 7: "The cat, the dog, the rabbit, the bear and the hamster walk to their favourite playground".

Spread 8: "The cat, the dog, the rabbit, the bear and the hamster play hide and seek until sundown. – Now I'm tired, said the cat".

POLLY: We said it was a mouse, the hamster. Who is this then with these long...?
    (points at the two long ears behind the bush)
ANDERS: The rabbit. I wonder where the others are?
POLLY: Where can they be? Now let's see. Who do we see? The cat. (points at the
    cat on the picture and unfolds one finger on her hand)
ANDERS: The rabbit and the hamster.

18

Spread 9: "Good night, said the hamster. Good night, said the bear. Good night, said the rabbit. Good night, said the dog. And good night, said the cat".

Polly points at the picture and unfolds two more fingers. Then there's two who we don't see! (points at the two folded fingers on her hand)

ANDERS: And one more.
POLLY: Who? Let's have a look (turns to the previous spread 7). We can see the cat, we see the rabbit with the long ears, we see the tail of the mouse, but the bear and the dog we cannot see (unfolds three fingers while talking about the characters they see, then turns back to spread 8 again). If you were the bear, where would you hide?
ANDERS: Out in the woods.
POLLY: In the woods. And if you were the dog, where would you hide then?
ANDERS: Just run away.
   Polly reads the following spread 9.
ANDERS: But now they're all back again! (turns back to spread 7). The dog and bear were missing.

The setting of the story is a game of hide-and-seek, which the child recognises immediately, knowing that some of the characters have run and hidden while the cat is fully visible as the "counter" in the game. There are some clues given in the picture, which connects the previous spread with the current one (the previous spread puts all the characters on display in a row, with their characteristics such as tails and ears visible to the reader). The teacher makes use of this connection, and mathematises the task of figuring out who is missing by using her fingers to keep track of which characters they see and which are hidden. The finger pattern she produces when keeping track is then used to structure the whole of five characters and the three visible characters, which allows the child to also see the relation between the visible and the hidden – the parts and the whole.

Discussion

Teaching in the dynamic pedagogical practice for which pre-school stands is not easily mastered. This, since the children in the pre-schools we have observed have very different experiences of mathematics, whereas the pre-school education is to support each individual child's potential for learning rather than a standard knowledge goal. Furthermore, pre-school includes children of different ages, which even more greatly widens the variation of competences in the field of mathematics that the children express in their interaction with teachers and peers.

MKT is without doubt as important for early childhood teachers as for any teacher at higher levels of education. The challenge for early childhood educators is perhaps that many mathematical notions that are basic and taken for granted once you have learnt them are new to young children, which is why it is difficult for an adult to understand a child's way of experiencing a notion. That is, it is hard to know what the child has not yet discerned. Furthermore, the teacher's pedagogical content knowledge is also put to the test if children are to be offered genuine opportunities to learn new meanings of mathematical notions and strategies, as the examples in this chapter have shown.

What more knowledge is there, then, that pre-school teachers need, in order to teach while reading a picture book? In our study, one feature of the interaction enacted in the book reading stands out, and this is characterised by flexibility: firstly, the flexibility to consider the children's individual competences and particularly to consider what aspects of a learning object a particular child may need to extend his or her understanding; and secondly, the flexibility to keep a learning object present while following the children's initiatives. Determining and exploring the potential of the children's initiatives in relation to a teaching goal demands knowledge not only of the mathematics and how the child perceives a specific notion, but also of how to direct the child's attention to necessary aspects of that notion. In VT terms, the teacher can offer a child opportunity to discern such aspects by contrasting its features. Teaching in this way constitutes a fusion of the different parts of the MKT that the teacher possesses.

The analysis illustrates how children's learning opportunities are closely related to the teachers' ways of making connections within and between mathematical notions in the reading sessions. Such connections can be made very subtly, but with significant effects on the learning opportunities (Ekdahl, 2019). Whether connections are made that widen the child's experiences of a mathematical notion, we suggest, depends on the teacher's knowledge of the specific content, in order to be able to note contrasts through illustrative examples and to avoid bringing in notions in isolation. In our study, we can see episodes of both kinds, in which the teacher supports concept development (see Watson & Mason, 2006) as the teacher and child interact around the same notion, as well as episodes in which many different notions are brought in with a lack of depth and exploration.

References

Ball, D. L., Thames, M. H., & Phelps, G. (2008). Content knowledge for teaching: What makes it special? *Journal of Teacher Education, 59*(5), 389–407.

Bennett, J., & Tayler, C. P. (2006). *Starting strong II. Early childhood education and care.* Paris: OECD.

Björklund, C., & Palmér, H. (2020). Preschoolers reasoning about numbers in picture books. *Mathematical Thinking and Learning, 22*(3), 195–213. https://doi.org/10.1080/10986065.2020.1741334

Blömeke, S., Gustafsson, J.-E., & Shavelson, R. J. (2015). Beyond dichotomies: Competence viewed as a continuum. *Zeitschrift für Psychologie, 223*(1), 3–13.

Butterfield, L. D., Borgen, W. A., Amundson, N. E., & Maglio, A.-S. (2005). Fifty years of the critical incident technique: 1954–2004 and beyond. *Qualitative Research, 5*(4), 475–497.

Davis, B., & Simmt, E. (2006). Mathematics-for-teaching: An ongoing investigation of the mathematics that teachers (need to) know. *Educational Studies in Mathematics, 61*(3), 293–319.

Ekdahl, A.-L. (2019). *Teaching for the learning of additive part-whole relations. The power of variation and connections.* (Diss.). Series no. 038/2019. Jönköping: Jönköping University, School of Education and Communication.

Greenhoot, A. F., Beyer, A. M., & Curtis, J. (2014). More than pretty pictures? How illustrations affect parent-child story reading and children's story recall. In J. S. Horst & C. Houston-Price (Eds.), *What and how young children learn from picture and story books.* Open Book: Frontiers in Psychology. https://doi.org/10.3389/fpsyg.2014.00738

Hassinger-Das, B., Jordan, N. C., & Dyson, N. (2015). Reading stories to learn math: Mathematics vocabulary instruction for children with early numeracy difficulties. *The Elementary School Journal, 116,* 242–264.

Jennings, C. M., Jennings, J. E., Richey, J., & Dixon-Krauss, L. (1992). Increasing interest and achievement in mathematics through children's literature. *Early Childhood Research Quarterly, 7,* 263–276.

Llinares, S. & Krainer, K. (2006). Mathematics (students) teachers and teacher educators as learners. In A. Guitérrez & P. Boero (Eds.), *Handbook of research on the psychology of mathematics education: Past, present and future* (pp. 429–459). Rotterdam: Sense Publishers.

Marton, F. (2015). *Necessary conditions of learning.* New York: Routledge.

Mosvold, R., Bjuland, R., Fauskanger, J., & Jakobsen, A. (2011). Similar but different – investigating the use of MKT in a Norwegian kindergarten setting. In M. Pytlak, T. Rowland & E. Swoboda (Eds.), *Proceeding of the seventh congress of the European society for research in mathematics education* (pp. 1802–1811). Rzeszów: University of Rzeszów.

National Agency for Education (2018). *Curriculum for the preschool. Lpfö 18.* Stockholm: National Agency for Education.

National Agency for Education (2019). *Barn och personal i förskolan per 15 oktober 2018* [Children and staff in preschool October 15th 2018]. No: 5.1.1-2019.321.

Pramling, N., & Pramling Samuelsson, I. (2008). Identifying and solving problems: Making sense of basic mathematics through storytelling in the preschool class. *International Journal of Early Childhood, 40*(1), 65–79.

Pramling, N., Wallerstedt, C., Lagerlöf, P., Björklund, C., Kultti, A., Palmér, H., Magnusson, M., Thulin, S., Jonsson, A., & Pramling Samuelsson, I. (2019). *Play-responsive teaching in early childhood education.* Dordrecht, the Netherlands: Springer.

Shiyan, O., Björklund, C., & Pramling Samuelsson, I. (2018). Narratives as tools for expressing structure and creativity with preschool children in two cultures. In N. Veraksa & S. Sheridan (Eds.), *Vygotsky's theory in early childhood education and research. Russian and western values* (pp. 38–53). New York, NY: Routledge.

Sowder, J. T. (2007). The mathematical education and development of teachers. In F. K. Lester (Ed.), *Second handbook of research on mathematics teaching and learning* (pp. 157–224). Charlotte: National Council of Teachers of Mathematics & Information Age Publishing.

Swedish Research Council (2017). *God forskningssed* [Good ethics in research]. Stockholm: The Swedish Research Council.

Van den Heuvel-Panhuizen, M., & Elia, I. (2012). Developing a framework for the evaluation of picture books that support kindergartners' learning of mathematics. *Research in Mathematics Education, 14*(1), 17–47.

Van den Heuvel-Panhuizen, M., & Elia, I. (2013). The role of picture books in young children's mathematics learning. In L. D. English & J. T. Mulligan (Eds.), *Reconceptualizing early mathematics learning* (pp. 227–251). Dordrecht: Springer.

Van den Heuvel-Panhuizen, M., & Van den Boogaard, S. (2008). Picture books as an impetus for kindergartners' mathematical thinking. *Mathematical Thinking and Learning, 10*(4), 341–373.

Ward, J. M., Mazzocco, M. M., Bock, A. M., & Prokes, N. A. (2017). Are content and structural features of counting books aligned with research on numeracy development? *Early Childhood Research Quarterly, 39*, 47–63.

Watson, A., & Mason, J. (2006). Variation and mathematical structure. *Mathematics Teaching, 194*, 3–5.

Wilson, S., & Cooney, T. (2002). Mathematics teacher change and development: The role of beliefs. In G. C. Leder, E. Pehkonen & G. Törner (Eds.), *Beliefs: A hidden variable in mathematics education?* (pp. 127–147). Dordrecht: Kluwer Academic Publishers.

Chapter 12

The relative frequency of mathematical learning opportunities in the morning circle in relation to the development of children's mathematical competencies

*Aljoscha Jegodtka, Georg Hosoya, Markus Szczesny,
Lars Jenßen, Corinna Schmude*

Introduction

The development of mathematical competencies in children is of central importance to exploring the world from a child's perspective, to their performance at school (Duncan et al., 2007; Vandell, Burchinal, & Pierce, 2016) and to their future life (OECD, 2018). The development of children's mathematical competencies depends on the one hand on their individuality (Krajewski, Nieding, & Schneider, 2008); on the other hand, the learning environment of the educational system plays a significant role (Lehrl, Kluczniok, & Rossbach, 2016). In a meta-analysis of early childhood education and care (ECEC) institutions in nine European countries, Ulferts, Wolf and Anders (2019) were able to establish that the domain-specific process quality is of particularly importance: "Domain-specific process quality […] seems to be of particular value to outcomes in mathematics, which indicates a stronger dependency on specific stimulation and promotion" (Ulferts et al., 2019: 1484). According to Kluczniok, Sechtig and Roßbach (2012, p. 34), process quality can be understood not only in terms of the interactions between children and early childhood educators, but also with respect to the room as a third educator and the material available.

In regards to the design of mathematical teaching-learning settings in ECEC institutions, there are two very different approaches that correspond to the two ECEC systems identified by the OECD (2006): the *social pedagogy* approach (e.g. Germany, Norway) and the *readiness for school* approach (e.g. UK and US). In Germany, the *social pedagogy* approach is more prominent. "*Sozialpädagogik* (social pedagogy) is a theory, practice and profession for working with children […]. The social approach is inherently holistic. The pedagogue sets out to address the whole child, the child with body, mind, emotions, creativity, history and social identity. […]. For the pedagogue, working with the whole child, learning, care and, more generally, upbringing (the elements of the

DOI: 10.4324/9781003172529-12

original German concept of pedagogy: *Bildung*, *Erziehung* and *Betreuung*) are closely related – indeed inseparable activities at the level of daily work. These are not separate fields needing to be joined up, but interconnected parts of the child's life" (OECD, 2004, p. 19). In German ECEC institutions, the pedagogy is primarily situated in the everyday pedagogical routine with regard to early mathematical education (e.g. Benz, Peter-Koop, & Grüßing, 2015; Fthenakis, Schmitt, Daut, Eitel, & Wendell, 2008). This anchoring of early mathematical education in everyday life implies that it is particularly important that the early childhood educator stimulates mathematical educational processes in a way that is integrated into everyday life.

It is well known that in open settings, sustained shared thinking is of special significance Siraj-Blatchford, Sylva, Muttock, Gilden, & Bell, 2002), and according to Wood, Bruner and Ross (1976), scaffolding plays a special role for children's cognitive development in situations that tend to be controlled. Furthermore, a distinction can be made between inclusive and exclusive modes of interaction (Jegodtka, Schmude, & Eilerts, 2019; Nentwig-Gesemann, Wedekind, Gerstenberg, & Tengler, 2012). This knowledge about the relevance of early childhood educator[1]-child interactions always refers to specific domains of early education (Pohle, Hosoya, Loftfield, & Jenßen, 2019).

Empirical studies located in the Anglo-American region (e.g. Clements & Sarama, 2011; Klibanoff, Levine, Huttenlocher, Vasilyeva, & Hedges, 2006; Pianta et al., 2014) emphasize the role of early childhood educators in the context of early mathematical education. It is important that the educator take up and develop the corresponding impulses of the children and encourage interactions with mathematical content (Pohle et al., 2019). Current studies (e.g. Schuler & Sturm, 2019) indicate that the active focus on mathematical content by early childhood educators in the framework of mathematical rule games clearly encourages process-related mathematical aspects. Overall, however, it must be stated that for the European region, and for Germany especially, examining the role of early childhood educators is a desideratum for research (European Commission, EACEA, Eurydice and Eurostat, 2014; Köller, 2016). In particular, studies that focus on the role of the individual early childhood educator in developing mathematical competencies in children on the individual level are not yet available. This chapter presents an explorative study that focuses on the relative frequency of mathematical learning opportunities between children and their respective childhood educator in the framework of the morning circle and what effect that relative frequency has on the individual development of the children's mathematical competence.

Research question

The theoretical framework of the *Pro-KomMa* project (Professionalization of Early Childhood Education Studies in Mathematics) examines the professional mathematical competence of early childhood educators who have only recently started their careers (Jenßen et al., 2016). With a view to mathematical dispositions, skills and outcomes, the goal is to trace the entire path from the early childhood educators' mathematical dispositions through their situation-specific skills to their performance and to the mathematical competencies (outcomes) that can be identified in children. In the present study, we are focusing on the interactions between early childhood educators and children. Based on the evidence available regarding the relevance of interactions between early childhood educators and children to the cognitive development of those children, it can be assumed that the frequency of mathematical interactions has an influence on the development of children's mathematical competence. Accordingly, the following research question must be answered: Is there a connection between the relative frequency of mathematical learning opportunities within the framework of the morning circle and the development of mathematical competencies in children?

Methodology

Morning circles were videotaped (see e.g. Fritzsche & Wagner-Willi, 2014) in various preschool institutions. The aim was to videotape each early childhood educator three times, but due to the specific nature of the field and the complexity of field access, this was not always possible. If, for example, an early childhood educator in the institution fell ill on the day of data collection, children were allocated into new groups; hence, not all children present had submitted the consent needed in order to include them in the videography.

Technical requirements and each decision made regarding camera operation always exert an influence on the data available. What was available for recording in the present case was a camera with a microphone attached. The technical characteristics of the camera (e.g. focal length of the lens) determined the portion of in-situ pedagogical activity able to be included in the respective video. The focus was on the early childhood educators, as their performance is of primary interest for the purposes of this analysis. Since their pedagogical actions were carried out in relation to the children, the pedagogical situation became a subject of consideration. If, for example, the early childhood educator and the children sat in a circle and bent over an object, the camera could not capture how they interacted with the object. In this respect, the context of the interaction must be considered when interpreting the videotaped situations. The video recordings were made by employees in the *Pro-KomMa* project. One employee at a time recorded a morning circle, holding the camera in hand and focusing on the early childhood educator. They stood as far outside the morning circle as technically possible so as to make sure it could be enacted under conditions that were as normal as possible.

In addition, the development of the mathematical competencies of the children supervised by the early childhood educators was measured and recorded at two different points in time. Examination was given to the correlation between the relative frequency of mathematical learning opportunities in the morning circle and the development of children's mathematical competencies.

The development of children's mathematical competencies was assessed with the MBK-0 (test targeting basic mathematical competencies at kindergarten age: Seeger, Holodynski, & Souvignier, 2014). Based on the modeled acquisition of quantity-number-word-linkage that can generally be expected to develop at an age of about 3.5 years and older, the MBK-0 provides information with regard to numerical competencies and their development in preschool-age children. As sub-dimensions of numerical competence, number words and numbers (without size reference), the quantity-number-word-linkage and number relations were taken into consideration. The MBK-0 has an internal consistency of Cronbach's alpha = .95 and a retest reliability (4–6 months) of r_{tt} = .83–.88 (Krajewski, 2018). To carry out the MBK-0 test, the participating educators were asked to randomly select ten children in their care, who were then tested and measured by trained observers at two distinct points separated by an interval of about six months. The MBK-0 test was developed for screening kindergarten children over longer intervals of time (e.g. annually or semi-annually). Thus, only two measurements were possible in the given test period.

Sample

The participating early childhood educators were recruited in Berlin and from multiple districts and urban municipalities in the federal state of Brandenburg. They worked in preschool institutions in small- and medium-sized towns within the federal state of Brandenburg and in various districts of Berlin. A total of $n = 11$ early childhood educators participated in the videography study, and the state of their children's mathematical development was measured at two points in time.

At the time of measurement, the average age of the participating early childhood educators was $M = 35.5$ years (the median was 29.5 years, ranging from 25 to 54 years). Two early childhood educators were male, nine were female. Two early childhood educators graduated from a university of applied sciences; all the other educators completed vocational training to qualify as early childhood educators. Four early childhood educators completed additional vocational training before working in early childhood education. All of the early childhood educators were first-time employees and within the first five years of their careers.[2]

A total of $N = 92$ children participated in the study (see Table 12.2), $N = 83$ at the first point of measurement and $N = 76$ at the second. The age of the children examined was $M = 4.2$ years at the first point of measurement (ranging from 2.7 to 5.5 years) and $M = 4.8$ years at the second (ranging from 3.1 to 6.2 years). At both measurement points, slightly more girls than boys were examined (with girls constituting 56% at the first point of measurement and 54% at the second point of measurement).

Available videography data

Table 12.1 represents the video material available for the 11 early childhood educators.

Table 12.1 Duration of the morning circles

Chiffre educator	Morning circle 1	Morning circle 2	Morning circle 3	Total duration of the morning circles
Educator 01	0:47:38	0:26:38	0:16:46	1:31:02
Educator 02	0:53:30	–	–	0:53:30
Educator 03	0:12:30	0:09:57	–	0:22:27
Educator 04	0:22:57	–	–	0:22:57
Educator 05	0:25:03	0:16:24	0:14:11	0:55:38
Educator 06	0:51:23	–	–	0:51:23
Educator 07	0:13:30	–	–	0:13:30
Educator 08	0:05:01	0:07:41	0:14:47	0:27:29
Educator 09	0:20:58	0:13:03	0:21:59	0:56:00
Educator 10	0:17:00	0:21:39	0:19:25	0:58:04
Educator 11	0:34:36	0:21:41	0:22:57	1:19:14

This means that a total of 24 videotaped morning circles are available for the analysis, although they differ considerably in length.

Data analysis

The video material was evaluated based on a coding manual (Mayring, 2000). This manual was initially developed in a multi-step process. The first step consisted of the development of deductive categories. In the second step, categories were inductively generated by means of category differentiation and expansion on the basis of 10% of the video material. The resulting coding manual was then discursively validated by experts, with four early childhood educators and five scientists from the fields of educational science (1×), mathematical didactics (1×) and childhood education (3×) participating in this panel of experts. Subsequently, external experts – one researcher from the field of childhood pedagogy and two early childhood educators – independently rated another 10% of the video material based on the consensual coding manual. In the final round, the same expert participants evaluated and discussed their experiences with the application of the manual and the results of the coding. The coding manual was then produced in its final version. The development process of the coding manual described here was also used to train the raters.

To rate the material, partial transcriptions of the videos were made according to simple rules. All relevant sequences, i.e. sequences with mathematical content, were transcribed. The videos were divided into 30-second sequences, and each sequence was coded individually, whereby the coding was always carried out twice according to type and content: Type refers to the mathematics-related behaviour of the early childhood educator, with the categories being (1) no mathematical impulse by the educator, (2) mathematical impulse (verbal or non-verbal) by the educator not followed by a reaction of the children and (3) mathematical interaction between educator and child(ren). Each sequence was coded with respect to the mathematical domain addressed (e.g. number or geometry).

Quality criteria

There are many criteria for judging quality in qualitative research. In the following, the ecological validity, the resulting data material and the work with external experts are evaluated and the inter-rater reliability presented.

Ecological validity, in the sense of the question as to what extent the situation being researched represents the reality being researched, is considered via three mechanisms: Firstly, the early childhood educators were asked to act as normal as possible in the situation. In other words, they were instructed not to prepare any special activities and to carry on as naturally as possible. Secondly, a self-description of the early childhood educators was collected; they were asked whether the fact that they were recorded had changed their behaviour, whether unusual situations had occurred and/or whether they had followed a program specifically for that day that deviated from their daily pedagogical routine. This was consistently met with a denial. Finally, the repeated videography of the three different morning circles per educator was intended to compensate for the special features of any given day – for example, the birthday of a child.

One difficulty regarding the data material is that the target number of videotaped morning circles per early childhood educator was not reached consistently. The reasons for this were, for instance, a work-overload-related discontinuation of participation in the study (which occurred once), field conditions such as the presence of children without the necessary consent for participation in the study, or the illness or pregnancy of early childhood educators.

The quality of data interpretation should be improved through collaborative work. This should expand the limitations of understanding that a single interpreter can have. For this purpose, expert rounds were held in order to include various possible perspectives on the available data (see Reichertz, 2013).

To ensure a high rating quality, 10% of all videographed data material was coded by two different coders, and the inter-rater reliability was estimated with Cohen's kappa (see e.g. Wirtz & Casper, 2002). Landis and Koch (1977) postulate that a Cohen's kappa value of .61–.80 is substantial and a Cohen's kappa value of .81–1.00 is approximately perfect. A percentage agreement of 87.9% and a Cohen's kappa of .76 were available (see e.g. Kuckartz, 2018).

Results

There were considerable differences in the duration of the respective morning circles among early childhood educators; the shortest circle lasted five minutes, while the longest was about 53 minutes and 30 seconds. The frequency of early childhood educator-child interactions with regard to mathematical content varied between nine and 113. For each early childhood educator, the recorded morning circles were divided into 30-second-long sequences. Each sequence was rated for the presence or absence of interactions with mathematical content

Table 12.2 Descriptive statics

Code educator	Percentage of interactions with mathematical content	Number of children	Number of children for whom a difference in MBK0 scores could be computed	Group mean difference in MBK0 scores (t^2-t^1)
Educator 01	62.08%	9	8	11.25
Educator 02	39.25%	9	4	10.25
Educator 03	33.33%	4	3	13.00
Educator 04	19.56%	10	6	5.50
Educator 05	13.51%	6	1	1.50
Educator 06	62.13%	10	9	11.00
Educator 07	48.14%	10	5	9.20
Educator 08	21.81%	5	2	7.00
Educator 09	45.96%	8	7	9.86
Educator 10	36.20%	10	7	6.57
Educator 11	31.01%	11	5	7.00

according to the coding rules. For instance, when in four of ten sequences mathematical content was present, the resulting percentage of interactions with mathematical content was 40%.

Table 12.2 shows the descriptive statistics. The percentage of interactions focused on mathematical content within a group ranged from min = 13.51% to max = 62.08%. The number of children per group ranged from min = 5 to max = 11. The difference in MBK0-scores could not be computed for all children, as test data were not available at both points of measurement. The range of children per group for which data are available was min = 1 to max = 9. The range of differences in MBK0-scores was min = 1.5 to max = 13.00, with an average group gain of $M = 8.38$ points, indicating on a descriptive level that overall, the groups' performance on the MBK0 improved between the testing occasions. To assess on an exploratory level whether the group gains in MBK0 scores were related to the percentage of interactions with mathematical content, we calculated the nonparametric rank-correlation-coefficient Kendall's tau. The coefficient ranges from –1, indicating a perfect negative association, to 1, indicating a perfect positive association. A Kendall's tau of 0 indicates that there is no systematic relationship between the two variables. In the present case, the estimated coefficient is tau = 0.57. The associated z-value of $z = 2.428$ ($p = .015$, two-tailed) indicates that it is relatively unlikely to detect such a high estimate of tau when, in fact, the null-hypothesis of no association in the population is true. Accordingly, the statistical analysis suggests a positive relationship between the relative frequency of interactions with mathematical content and the children's competence growth on the group level. Figure 12.1 illustrates this relationship.

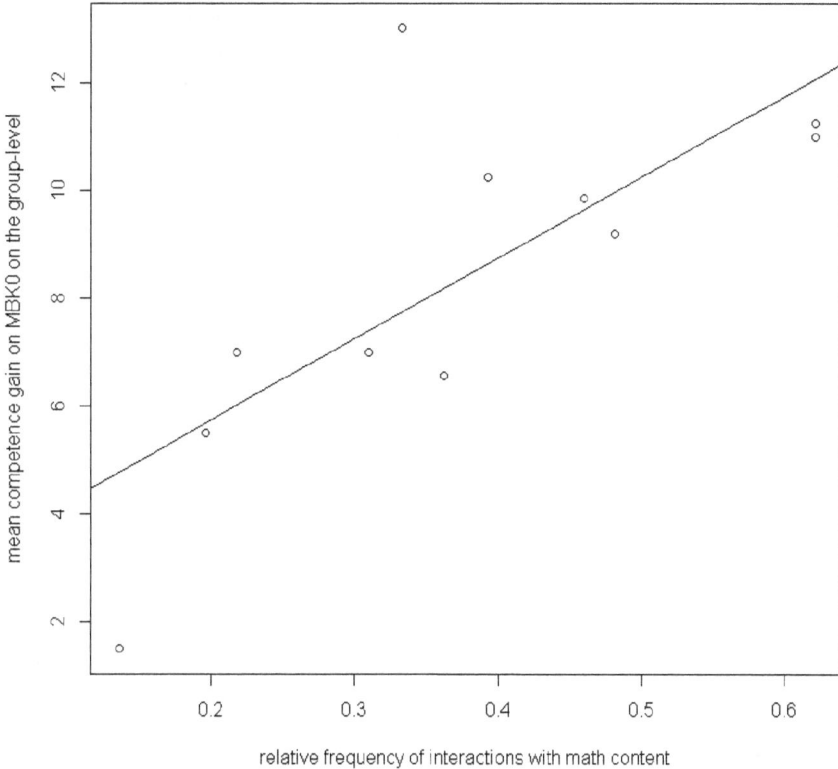

Figure 12.1 Scatterplot of the relative frequency of interactions of mathematical content versus mean competence gain on the MBK0 on the group level.

Note. The line is based on a linear regression analysis for visualization purposes. The data used can be found in Table 12.2 (columns labeled "Percentage of interactions with mathematical content" and "Group mean difference in MBK0 scores"). On the x-axis, percentages were scaled to range between 0 (0%) and 1 (100%). The strength of the association was assessed using Kendall's tau (see text).

Conclusion and discussion

It is known that there is a positive correlation between a child's cognitive development and the child's attendance at preschool institutions (Ulferts & Anders, 2016) and that process quality plays a particularly important role in that respect (Ulferts et al., 2019). This exploratory study was able to establish a positive correlation linking the relative frequency of mathematical interactions between early childhood educators and children in the morning circle setting and the development of mathematical competencies in children.

There are also, however, some underlying limitations. First, it must be taken into consideration that the sample of early childhood educators is relatively small.

It would be desirable to replicate this result with a larger and different sample to check whether the result is merely an idiosyncrasy of the present data set or indeed a substantively meaningful relationship between the early childhood educators' behaviour in their respective groups and the children's competence gain on the group level. Second, it would be advisable to account for the missing children's test data within a more advanced analytical framework. For future studies, it would be beneficial to attune the sampling procedures to the analytical methods and to run the study with enough power to detect possible positive associations with parametric methods, such as multilevel modeling, for example. Furthermore, the sample suggests a certain degree of bias; the participants were aware of the fact that early mathematical education was under assessment – and thus, accordingly, a certain openness towards this topic can be assumed, which may be associated with a stronger focus on mathematical content during the videotaping procedure.

Another central challenge to the research originates in the field of ECEC itself: There is no standardized group constellation. For instance, the group size per early childhood educator as well as the qualification of the early childhood educator may vary (see e.g. Oberhuemer & Schreyer, 2017). In addition, the organization of daily routines as well as the total interaction time between early childhood educator and children is very heterogeneous across the German ECEC system. The freedom provided to educators in structuring the daily routine also plays a major role. As became clearly evident, the duration of the morning circles differs considerably. In this respect, the relative frequency of the mathematical learning opportunities can only be compared to a limited extent. However, it can be assumed that the mathematical interactions made possible by early childhood educators do not only take place in the morning circle but also in everyday preschool life. That this assumption is not entirely off base can be concluded from the fact that the development of the children's mathematical skills does not depend on the length of the morning circle. Overall, it is difficult to generalize the results presented here. Apart from a larger sample size, it would be beneficial to collect data about the various institutional conditions and to categorize them (e.g. group size, open work without fixed group constellations etc.). Last, but not least: It could be shown previously (Jegodtka et al., 2019) that there are great differences in the quality of the design of interactions with mathematical content. Accordingly, future research should also consider quality of interaction as an aspect that may well influence the development of children's mathematical skills.

Despite these limitations, some consequences can be drawn for practical work in ECEC institutions: First, it seems important for early childhood educators to regularly address mathematical aspects in their interactions with children in order to promote the development of their protégés' mathematical competencies. Within the framework of the pedagogical approach in Germany, this primarily means recognizing mathematical moments in everyday situations in order to bring them up in interactions with children. Thus, it is important to "design opportunities to learn mathematics or (...) use situations spontaneously for mathematical learning by asking adequate questions or choosing appropriate learning material" (Gasteiger & Benz, 2018, p. 112). Second, empirical findings indicate (Schuler & Sturm, 2019) that this also applies to mathematical rule games: The deliberate linguistic support of the play activities promotes the process-related skills of children.

Notes

1 For the various professions and job titles in German ECEC settings, see also Oberhuemer and Schreyer (2017).
2 In order to be allowed to work as an early childhood educator in Germany, vocational training or a university degree as an "educator" or as a "childhood educator" must be completed with state recognition. See also Oberhuemer and Schreyer (2017).

References

Benz, C., Peter-Koop, A., & Grüßing, M. (2015). *Frühe mathematische Bildung. Mathematiklernen der Drei- bis Achtjährigen*. Berlin/Heidelberg: Springer Spektrum.

Clements, D. H., & Sarama, J. (2011). Early childhood mathematics intervention. *Science, 333*, 968–970. http://dx.doi.org/10.1126/science.1204537.

Duncan, G. J., Dowsett, C. J., Claessens, A., Magnuson, K., Huston, A. C., Klebanov, P., & Japel, C. (2007). School readiness and later achievement. *Developmental Psychology, 43*(6), 1428–1446. http://dx.doi.org/10.1037/0012-1649.43.6.1428

European Commission, EACEA, Eurydice, & Eurostat (2014). *Key data on early childhood education and care in Europe. 2014 edition. Eurydice and Eurostat report*. Luxembourg: Publications Office of the European Union.

Fritzsche, B. & Wagner-Willi, M. (2014). Dokumentarische Interpretation von Unterrichtsvideografien. In R. Bohnsack, B. Fritzsche & M. Wagner-Will (Hrsg.), *Dokumentarische Video- und Filminterpretation. Methodologie und Forschungspraxis* (pp. 131–152) Opladen: Barbara Budrich.

Fthenakis, W. E., Schmitt, A., Daut, M., Eitel A., & Wendell, A. (2008): *Frühe mathematische Bildung. Natur-Wissen schaffen 2*. Essen/Köln: Bildungsverlag EINS.

Gasteiger, H., & Benz, C. (2018). Enhancing and analyzing kindergarten teachers' professional knowledge for early mathematics education. *Journal of Mathematical Behavior, 51*, 109–117. https://doi.org/10.1016/j.jmathb.2018.01.002

Jegodtka, A., Schmude, C., & Eilerts, K. (2019). Exkludierende und inkludierende mathematikhaltige Interaktionsmodi zwischen frühpädagogischer Fachkraft und Kind(ern) in typischen Alltagssituationen in Kindertagesstätten. In D. Weltzien, H. Wadepohl, C. Schmude, H. Wedekind, & A. Jegodtka, *Forschung in der Frühpädagogik. Band XII: Schwerpunkt Interaktionen und Settings in der frühen MINT-Bildung* (pp. 31–57). Freiburg i.Br.: FEL-Verlag.

Jenßen, L., Jegodtka, A., Eilerts, K., Eid, M., Koinzer, T., Schmude, C., Rasche, J., Szczesny, M., & Blömeke, S. (2016). Pro-KomMa – Professionalization of early childhood teacher education: Convergent, discriminant, and prognostic validation of the KomMa models and tests. In H. A. Pant, O. Zlatkin-Troitschanskaia, C. Lautenbach, M. Toepper, & D. Molerov (Eds.), *Modeling and measuring competencies in higher education – Validation and methodological innovations (KoKoHs) – Overview of the research projects* (pp. 39–43). Berlin & Mainz: Humboldt University & Johannes Gutenberg University.

Klibanoff, R. S., Levine, S. C., Huttenlocher, J., Vasilyeva, M., & Hedges, L. V. (2006). Preschool children's mathematical knowledge: The effect of teacher 'math talk'. *Developmental Psychology*, 42(1), 59–69. http://dx.doi.org/10.1037/0012-1649.42.1.59

Kluczniok, K., Sechtig, J., & Roßbach, H. G. (2012). Qualität im Kindergarten. Wie gut ist das Niveau der Kindertagesbetreuung in Deutschland und wie wird es gemessen? *DJI Impulse Das Bulletin des Deutschen Jugendinstituts*, 2(2012), 33–36.

Köller, O. (2016). Frühe Bildung als Herausforderung psychologischer Forschung: Editorial. *Psychologie in Erziehung und Unterricht*, 63, 1–2. http://dx.doi.org/10.2378/peu2016.art01d

Krajewski, K. (2018): *MBK-0. Test mathematischer Basiskompetenzen im Kindergartenalter.* Göttingen: Hogrefe.

Krajewski, K., Nieding, G., & Schneider, W. (2008). Kurz- und langfristige Effekte mathematischer Frühförderung im Kindergarten durch das Programm Menge, zählen, Zahlen. *Zeitschrift für Entwicklungspsychologie und Pädagogische Psychologie*, 40(3), 135–146.

Kuckartz, U. (2018). *Qualitative Inhaltsanalyse. Methoden, Praxis, Computerunterstützung* (4 Auflage). Weinheim & Basel: BeltzJuventa.

Landis, J., & Koch, G. (1977). The measurement of observer agreement for categorical data. *Biometrics*, 33(1), 159–174. doi:10.2307/2529310

Lehrl, S., Kluczniok, K., & Rossbach, H.-G. (2016). Longer-term associations of preschool education: The predictive role of preschool quality for the development of mathematical skills through elementary school. *Early Childhood Research Quarterly*, 36, 475–488.

Mayring, P. (2000). Qualitative content analysis [28 paragraphs]. Forum qualitative sozialforschung/forum: Qualitative social research, 1(2), Art. 20. http://nbn-resolving.de/urn:nbn:de:0114-fqs0002204 (accessed at 15.02.2021).

Nentwig-Gesemann, I., Wedekind, H., Gerstenberg. F., & Tengler, M. (2012). Die vielen Facetten des 'Forschens'. Eine ethnografische Studie zu Praktiken von Kindern und PädagogInnen im Rahmen eines naturwissenschaftlichen Bildungsangebots. In I. Nentwig-Gesemann, K. Fröhlich-Gildhoff, & H. Wedekind (Hrsg.), *Forschung in der Frühpädagogik V. Schwerpunkt: Naturwissenschaftliche Bildung – Begegnungen mit Dingen und Phänomenen* (pp. 33–64). Freiburg: FEL-Verlag.

Oberhuemer, P. & Schreyer, I. (2017). Germany – ECEC workforce profile. In P. Oberhuemer & I. Schreyer (Eds.), *Workforce profiles in systems of early childhood education and care in Europe*, www.seepro.eu/English/Country_Reports.htm (accessed at 15.02.2021).

OECD (2004). *OECD Country note. Early childhood education and care policy in the Federal Republic of Germany*. 26 November 2004. https://www.oecd.org/germany/33978768.pdf (accessed at 15.02.2021).

OECD (2006). *Starting strong II: Early childhood education and care.* http://www.oecd.org/education/school/startingstrongiiearlychildhoodeducationandcare.htm#ES (accessed at 15.02.2021).

OECD (2018). *Early learning matters: The international learning and well-being study.* Retrieved from http://www.oecd.org/education/school/Early-Learning-Matters-Project-Brochure.pdf (accessed at 15.02.2021).

Pianta, R., DeCoster, J., Cabell, S., Burchinal, M., Hamre, B., Downer, J., LoCasale-Crouch, J., Williford, A., & Howes, C. (2014). Dose–response relations between preschool teachers' exposure to components of professional development and increases in quality of their interactions with children. *Early Childhood Research Quarterly*, 29(4), 499–508. http://dx.doi.org/10.1016/j.ecresq.2014.06.001.

Pohle, L., Hosoya, G., Loftfield, C., & Jenßen, L. (2019). Indicators measuring preschool teachers' stimulation quality: Theoretical background and empirical testing. *Zeitschrift für Pädagogik, 65*(4), 525–541.

Reichertz, J. (2013). *Gemeinsam interpretieren. Die Gruppeninterpretation als kommunikativer Prozess.* Wiesbaden: VS Verlag für Sozialwissenschaften.

Schuler, S. & Sturm, N. (2019). Mathematische Aktivitäten von fünf- bis sechsjährigen Kindern beim spielen mathematischer Spiele – Lerngelegenheiten bei direkten und indirekten Formen der Unterstützung. In D. Weltzien, H. Wadepohl, C. Schmude, H. Wedekind, & A. Jegodtka, *Forschung in der Frühpädagogik. Band XII: Schwerpunkt Interaktionen und Settings in der frühen MINT-Bildung* (pp. 59–86). Freiburg: FEL-Verlag.

Seeger, D., Holodynski, M., & Souvignier, E. (2014). *BIKO-Screening zur Entwicklung von Basiskompetenzen für 3- bis 6-Jährige.* Göttingen: Hogrefe.

Siraj-Blatchford, I., Sylva, K., Muttock, S., Gilden, R., & Bell, D. (2002). *Researching effective pedagogy in the early years (REPEY).* London: Department for Education and Skills/Institute of Education, University of London.

Ulferts, H., & Anders, Y. (2016). Effects of ECEC on academic outcomes in literacy and mathematics: Meta-analysis of European longitudinal studies (CARE curriculum quality analysis and impact review of European (ECE), D4. 2. http://ecec-care.org/fileadmin/careproject/Publications/reports/CARE_WP4_D4_2_Metaanalysis_public.pdf (accessed at 15.02.2021).

Ulferts, H., Wolf, K. M., & Anders, Y. (2019). Impact of process quality in early childhood education and care on academic outcomes: Longitudinal meta-analysis. *Child Development, 90*(5), 1474–1489. https://doi.org/10.1111/cdev.13296

Vandell, D. L., Burchinal, M., & Pierce, K. M. (2016). Early child care and adolescent functioning at the end of high school: Results from the NICHD Study of Early Child Care and Youth Development. *Developmental Psychology, 52*(10), 1634–1654.

Wirtz, M. & Caspar, F. (2002). *Beurteilerübereinstimmung und Beurteilerreliabilität.* Göttingen: Hogrefe.

Wood, D. J., Bruner, J. S., & Ross, G. (1976). The role of tutoring in problem solving. *Journal of Child Psychiatry and Psychology, 17,* 89–100. http://dx.doi.org/10.1111/j.1469-7610.1976.tb00381.x

Chapter 13

Outlook: Context and its consequences

A neglected factor in research on early childhood teachers' professional skills?

Esther Brunner

Introduction

Who do we have in mind when we talk about early childhood teachers and their professional skills? And how is the professional field of work shaped for which they should acquire professional skills? These questions are all the more urgent because the job description of early childhood teachers does not relate to a standardized field, as is the case with teaching in school, for example. Despite considerable differences, models of professional knowledge of teachers are usually tailored to the context of schoolteachers and thus to a standardized field of work and action. The question is therefore whether common models can be transferred to the situation of early childhood teachers and, if so, to what extent they might be in need of adaptation.

Teaching—irrespective of a particular school level—is often, especially in the German-speaking research community, conceived of as an opportunity-usage structure ("Angebots-Nutzungs-Modell") (Figure 13.1) (Fend, 1998; Reusser & Pauli, 2010; Seidel, 2020).[1] This widely accepted modeling of teaching assumes that instruction has a social-interactive character and can therefore be interpreted as an interaction between the available opportunities to learn as provided by the teacher and their usage by the pupils. These opportunities to learn may differ in terms of goals, methods, or content. The basic assumption of such models is that an opportunity in itself is not sufficient to achieve learning success, but that an active use of the opportunities on the part of the learners is necessary to bring about an effect. Instruction is therefore fundamentally understood as an interconnected structure of teaching and learning processes.

The central task of the teacher in such a structure consists both in the planning, preparation, and provision of suitable opportunities to learn and in the active lending of learning support during the enactment of the lessons. In order to perform this task, teachers need planning and reflection skills as well as action-related skills (Lindmeier, 2011)[2] that are grounded in solid content knowledge (CK, here MCK), pedagogical content knowledge (PCK, here MPCK), and general pedagogical knowledge (GPK).

DOI: 10.4324/9781003172529-13

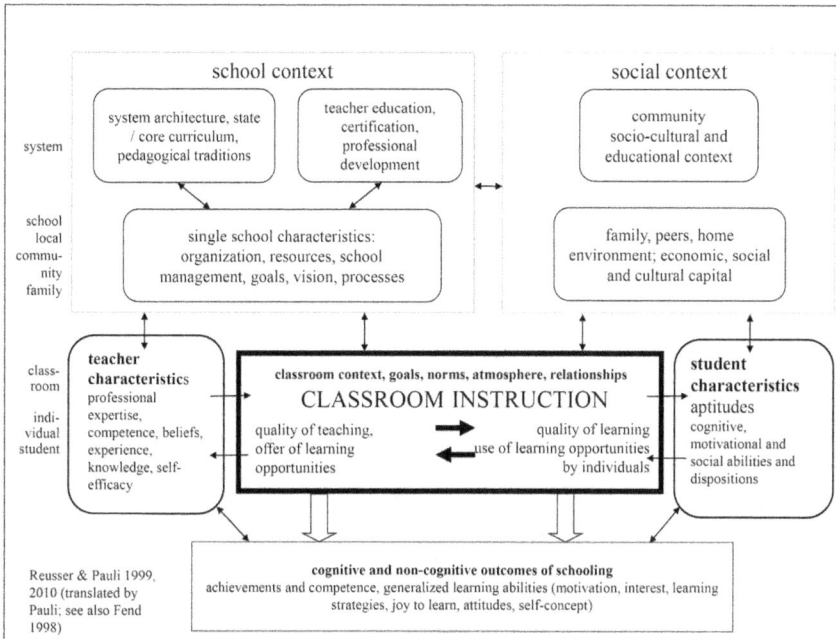

Figure 13.1 "Model of the provision and uptake of learning opportunities" (Reusser & Pauli, 2010, p. 18, translated by Reusser and Pauli, reprinted with kind permission of the authors).

The central interactants in this model of teaching and learning are the teacher on the one hand and the learners on the other hand. Their individual scope for active participation is defined by their cognitive and motivational-affective dispositions as well as by their situation-specific skills and prior knowledge. The restrictions and possibilities are not only related to the personal level, however (Figure 13.1). Rather, teaching with its interactants is embedded in various framework conditions and context variables—both on the side of the teachers and on the side of the learners—and limited by the societal framework as a space of possibility. What aspects of the subject matter are dealt with and negotiated in the interaction between the opportunity and its use in the classroom thus depends on cultural values, curricular guidelines, convictions, attitudes, and personal characteristics of the participants.

Cultural and societal influences determine pedagogical practice in a merely rudimentarily standardized field of work much more than in a highly standardized field of action. Educational activities in the field of early childhood are therefore much more diverse and strongly influenced by societal and cultural norms than one might think (e.g., Gasteiger, Brunner, & Chen, 2021; Hammer & He, 2016). Given this assumption, the question arises as to which postulates shape teaching in the field of early education and to what extent common models of professional knowledge and professional competence of teachers need to be adapted to the profession of early childhood teachers.

These questions are addressed in this contribution. First, it examines and discusses selected practices of early mathematics education with respect to their specificity. This leads to a characterization of the professional group that enacts these practices and to the question as to what skills are necessary for doing this in a competent and effective way. These reflections are summarized and discussed against the background of the model of transforming professional competence (Blömeke, Gustafsson, & Shavelson, 2015).

Early mathematics education

A rudimentarily standardized field

Teaching in schools, for example, mathematics instruction, is regulated by curricula and educational standards (e.g., Common Core State Standards Initiative, 2012; D-EDK, 2014; KMK, 2005; NCTM, 2000), by teaching aids such as visualizations and illustrative tools and models (e.g., Giaquinto, Mancosu, Jorgensen, & Pedersen, 2005), and by textbooks. The latter are the central instrument for introducing and presenting the range of tasks that needs to be completed, particularly in mathematics instruction (Rezat, 2013; Valverde, Bianchi, Wolfe, Schmidt, & Houang, 2002). They are based on the prescribed educational standards, which, in turn, are an expression of cultural and societal norms with respect to subject-specific contents and education and, moreover, must comply with the norms of the academic discipline (e.g., Burton, 2009).

Standardization of teaching means that it is possible to describe in relatively clear terms what skills and facets of competence teachers need to acquire and refine in order to be able to act competently within the opportunity-use structure of teaching (e.g., Brühwiler & Blatchford, 2011; Seidel, 2014). If, by contrast, there is no or merely rudimentary standardization of teaching, it is first necessary to determine what characteristics specify teaching in a particular societal context. Keeping to the example of early mathematics education, this prepares the ground for a precise determination of the professional skills of early childhood teachers.

Foundations of early mathematics education: two stances

On the one hand, early mathematics education is tailored to the developmental and psychological capabilities of young children (e.g., Sarama & Clements, 2009) and, on the other hand, builds on central objectives of kindergarten as an educational level (e.g., Clements, Sarama, & DiBiase, 2004). These objectives are mostly normative in nature and thus strongly dependent on the societal and cultural context, which, in turn, leads to different rationales for early mathematics education. In this regard, there are two main (normative) stances: (1)

kindergarten as a social-pedagogical institution with an educational mission and (2) the concept of "readiness for school" (Jegodtka, Hosoya, Szczesny, Jenßen, & Schmude, 2022). Countries such as Germany, the Netherlands, and Switzerland tend to follow the first principle, while the United States is more strongly oriented toward the second principle (Pohle, Jenßen, & Eilerts, 2022; Rimm-Kaufman & Sandilos, 2017). These different focuses on early mathematical education, which can also occur in hybrid forms and with varying weight, lead to different educational programs, implementations, and design elements and may thus also place different demands on early childhood teachers with respect to the requisite professional skills.

In the context of early mathematics education, the educational mediation between the opportunities and their use often oscillates between instruction and construction (Presmeg, 2014, p. 9). Quite often, there is a tendency to focus on construction and, at the same time, to be reluctant to give instructions (e.g., Björklund & Palmer, 2022), which corresponds to the first stance— kindergarten as a social-pedagogical institution with an educational mission. In countries such as the United States, where early mathematics education tends to follow the second stance "readiness for school", the principle of direct instruction is usually more explicit (De Haan, Elbers, & Leseman, 2014) and early childhood teachers interpret their work as consisting in direct teaching (Chiatovich & Stipek, 2016; Pohle et al., 2022). Depending on the orientation of the profile, early childhood teachers need to possess partly different professional skills.

Against this background, Chen, McCray, Adams, and Leow (2014) argue for a middle way. They call this "intentional teaching" and recommend an integration of both instruction and construction to ensure a high quality in early mathematics education.

Rationale I: central approaches and necessary skills

The first stance on early mathematics education is grounded in several age-specific approaches that make reference to child-centered approaches (OECD, 2011), such as play. Play is considered to be of great importance in developmental psychology (e.g., Piaget, 1975; Rubin & Pepler, 1982) as well as in early childhood education (e.g., Bruce, 1991; Ciolan, 2013). It therefore makes sense to use play also for the purpose of early mathematics learning (e.g., van Oers, 2010), for example, by including pedagogically suitable games (e.g., Stebler, Vogt, Wolf, Hauser, & Rechsteiner, 2013). In practice, such mathematics-related games have not yet become a regular part of pedagogical concepts, however, and their potential has not been fully recognized so far (Pohle et al., 2022).

A second approach of early mathematics education promotes learning in "natural" situations (Gasteiger, 2012). As in early mathematics learning through play, learning in natural and everyday contexts is highly situational and takes place in informal settings (Gasteiger & Benz, 2018a, 2018b). The teacher takes advantage of such situations and enriches them by emphasizing the mathematical aspects and linking them to the children's previous experiences and their prior (informal) knowledge. This provides children with the opportunity to acquire mathematical knowledge although there is no explicit or only little scaffolding (Collins, Brown, & Newman, 1989).

A third approach of early mathematics education follows a more formal approach and suggests learning with and from stories and picture books (Ginsburg, 2022, in this volume), which is characterized by a higher degree of instruction-like and -guided teaching. The teacher can either make use of the (implicit) mathematical content in an already existing story (e.g., van den Boogaard, van den Heuvel-Panhuizen, & Scherer, 2007) or refer to a picture book that was explicitly designed for the conveying of mathematical contents (e.g., Björklund & Palmer, 2022).

Due to the often highly spontaneous nature of such activities and the need to recognize, challenge, and support mathematically rich activities "on the spot", the focus of research into this field is particularly directed toward the situational skills of early childhood teachers. These skills do not develop exclusively on the basis of cognitive and affective-motivational dispositions (Blömeke et al., 2015), however, but also rest on both explicit and implicit knowledge (Gasteiger & Benz, 2018a, 2018b), incorporate experience and relate to pedagogical action (Lindmeier, 2011).

Recognizing the mathematical content in play situations requires, in particular, appropriate CK. Several recent studies (Bruns, Gasteiger & Strahl, 2021; Bruns, Strahl, & Gasteiger, 2020) indicate that common models of professional competence seem to underestimate the importance of implicit and practical knowledge as a contribution to competent action. Bruns et al. (2021) therefore operationalize mathematics-related knowledge of early childhood educators as a continuum between the poles of theory/science and practice and distinguish between concepts of MCK that are markedly oriented toward the academic discipline ("science-related") and concepts of MCK that refer to immediate practice ("practice-related"). Since mathematical expertise influences MPCK as well as the ability of early childhood teachers to perceive situations that are potentially conducive to learning (e.g., Dunekacke, Jenßen, Eilerts, & Blömeke, 2016), it is important to ask what kind of mathematical expertise this exactly is and which components can be deemed particularly important with regard to professional skills.

As illustrated by the picture-book approach (van den Heuvel-Panhuizen & van den Boogaard, 2008), play or opportunities to learn in natural situations are not always informal or incidental, but can be planned, prepared, and intentionally stimulated, analogous to school-based learning. Such approaches, which do not necessitate immediate action under pressure (Wahl, 1991), require not only situation-specific skills but also reflection skills (Lindmeier, 2011), and rest on explicit knowledge that is possibly combined with implicit knowledge. It would therefore be worth investigating more closely the extent to which implicit knowledge is used in the transformation of professional competence into performance and under what circumstances lesson preparation can combine implicit knowledge with explicit knowledge so as to ensure successful adaptive pedagogical action.

Rationale II: central approaches and necessary skills

If, by contrast, early mathematics education is primarily conceived of as "readiness for school", the contents are also based on normative principles, but these relate to educational standards of the next education level and the preparation for their achievement rather than social-pedagogical development. This is why content-related pre-concepts, for example, in the area of counting or quantity awareness (e.g., Krajewski & Schneider, 2009), which are often imparted and dealt with by means of learning programs (e.g., Krajewski, Nieding, & Schneider, 2007) or teaching materials that have specifically been designed for the purpose of early learning (e.g., Wittmann & Müller, 2010), have received increased attention in recent years.

In comparison to the approaches that are characteristic of the first stance, this explicit orientation toward school standards also leads to a higher degree of standardization of early education. The concept of "readiness for school" focuses on school-like learning, ensures the provision of preparatory activities, anticipates follow-up activities, and transfers existing school-based educational standards to the area of early education. According to this rationale, the professional field of early childhood teachers comparable to that of teachers in general.

Teaching in accordance with Stance II requires early childhood teachers to base their practice explicitly on MCK and MPCK. In combination with reflection skills (Lindmeier, 2011), this subject-specific knowledge allows them to plan and enact their lessons in a way that is conducive to the children's acquisition of elementary mathematical concepts. Several models have refined the concept of MPCK by splitting it up into components such as explanatory knowledge, diagnostic knowledge in the sense of knowledge about the mathematical thinking of learners, and knowledge about mathematical problems (Baumert & Kunter, 2013). Knowledge about mathematical problems is particularly relevant in school teaching where task completion is the central formative element of mathematics education (Rezat, 2013; Valverde et al., 2002). With regard to professional skills of early mathematics education, Sarama, Clements, and Stark Guss (2022, p. 164) suggest the three skills, "understanding the goals", "understanding children's thinking and learning", and "understanding effective teaching". These skills relate to different and multiple areas of knowledge (Shulman, 1986) and do not exclusively refer to MPCK, however. Analogous to the model proposed by Baumert and Kunter (2013), knowledge about specific age-appropriate learning settings and learning opportunities could be added as a further component of professional competence.

Framework and general conditions of teaching

Which of the two stances dominates in early mathematics education in a given country depends on the cultural and societal context. This context not only encompasses the educational settings but also the children since they are "part of their culture and the context created by their culture" so that "they will engage in mathematical thinking that is generated from these" (Cooke & Jay, 2022, p. 143). This assumption is of special significance in the area of early childhood education because, as set forth in Section 2.1, there is a lack of standardization of content or subject matter at this educational level, which is why factors that relate to cultural influences are receiving increasingly more attention in research (Gasteiger et al., 2021; Hammer & He, 2016; Oberhuemer, 2005).

If early mathematics education is not regarded as a standardized field and the teaching practices in a country are shaped by its cultural values and norms, it cannot be assumed that early mathematics education requires teachers to possess the same professional skills or even the same facets of these skills across all national contexts. In other words, it seems likely that professional contexts in which early childhood teachers work may vary considerably. In order to illustrate this conclusion, the following section presents the framework conditions of early mathematical education in the countries in which the authors of this volume are active as examples.

Early childhood teachers: who are they?

A special group of teachers

Due to the lack of or only rudimentary standardization of the educational level and their training, which is considerably influenced by the cultural and societal context, it seems plausible to assume that the profile of early childhood teachers as a professional group varies remarkably across countries (Gasteiger et al., 2021). Moreover, they can also be assumed to differ significantly from schoolteachers in terms of personal characteristics, for example, in their relationship to the subject of mathematics, for which there is no curriculum in various countries (Table 13.1).

Early childhood teachers often appear to be "math-avoidant" teachers, especially since anxiety about mathematics has been shown to be "a factor in the career choice of prospective early childhood teachers" (Jenßen, 2022, p. 90). Interest in mathematics, by contrast, does not seem to be a determining factor in the decision to take up the profession, especially when early childhood teachers are trained as generalists (Jenßen, 2022). Although their emotions about mathematics are not as negative as they are often supposed and reported to be (Chen, McCray, Adams, & Leow, 2014), and negative emotions are not as strong as commonly suspected, there is no doubt that a teacher's emotions about mathematics matter and that negative emotions such as anxiety about mathematics are widespread (e.g., Thiel & Jenßen, 2018). This is problematic because

Table 13.1 Overview of framework conditions of early childhood education in selected countries

	USA	Australia	Germany	Austria	Switzerland (German-speaking part)	Belgium (Flanders)	Norway	Sweden
Age of children	3–6 years	4–5 years	3–6 years	3–6 years (pre-school/kindergarten) 0–3 nursery	4.5–6.5 years	2.5–6 years	1–5 years	1–6 years
Compulsory for children	Not compulsory, but most children attend	Not compulsory	Not compulsory	Last year of preschool/kindergarten is compulsory	Yes	Not compulsory, but 97% attend	Not compulsory, but 92% of the 1- to 5-year olds and 97% of the 3- to 5-year olds attend	Not compulsory, for 6-year olds compulsory
Part of official education system	Kindergarten yes and sometimes pre-K (4's)	Yes, in Western Australia (this differs in other states)	No	No	Yes	Yes	Not part of the education system, but it is usually considered to be a part of the education system, even though it is not compulsory	Yes

(Continued)

Table 13.1 (Continued)

	USA	Australia	Germany	Austria	Switzerland (German-speaking part)	Belgium (Flanders)	Norway	Sweden
Responsibility to run ECE	Varies	Government, but church-based schools also offer ECE classes	State for a number of structural aspects (e.g., child-teacher ratio); care providers (e.g., community, church, welfare association) for pedagogical aspects (additional "Bildungspläne" from the federal states)	Federal states of Austria (e.g., federal state of Styria, federal state of Carinthia, etc.) are responsible for legislation and implementation, ECE maintained by municipalities, regional or private providers, and church	State and municipality share responsibility	Flemish government decides on the curricular objectives and educational laws	State and municipality share responsibility	N.A.

(Continued)

	USA	Australia	Germany	Austria	Switzerland (German-speaking part)	Belgium (Flanders)	Norway	Sweden
Kind of initial education	BA, when part of official school system	4 years, university-trained teacher	Majority graduates from vocational schools; only a few university graduates (BA child pedagogy or other BA)	Diploma, 5-year upper-secondary school with vocational education (teacher training colleges for early childhood education—BafEPs) or four-semester post-secondary or five-/six-semester part-time vocational education and training courses at the teacher training college for early childhood education	BA, university of teacher education	BA, non-academic teacher training institute	BA, university, or university college	BA university degree, vocational degree
Job designation	Varies, depending on placement	EC teacher	Erzieher*in ("educator")	Kindergarten teacher	Kindergarten teacher	Kindergarten teacher	Kindergarten teacher	Preschool teacher

(Continued)

Table 13.1 (Continued)

	USA	Australia	Germany	Austria	Switzerland (German-speaking part)	Belgium (Flanders)	Norway	Sweden
Generalist—subject-specific	Generalist	Generalist	Generalist	Generalist	Generalist	Generalist	Generalist	Generalist
MPCK, MCK part of curriculum of initial education	Yes, but slightly	Yes	Depends on the federal state and the institution where the education took place	No	Yes	MPCK generally is, MCK usually not—but teacher training institutes are free to decide for themselves on the curriculum they offer	Yes	Yes
Mandatory mathematics curriculum for ECE	No	Yes, for the compulsory years	No; most "Bildungspläne" include mathematics-related aspects	No	Yes	No	Yes	Yes, but no prescribed achievement goals (only to strive toward)
Literacy-based or inner-mathematical	Literacy-based	Balanced	Depends on the respective "Bildungsplan"	—	Literacy-based	N.A.	Literacy-based	N.A.

(Continued)

	USA	Australia	Germany	Austria	Switzerland (German-speaking part)	Belgium (Flanders)	Norway	Sweden
Parents' estimation, expectations	Play-based learning—some math, mostly social-emotional	N.A.	Depends on several aspects, groups of parents would deem it necessary, especially in the last year before school begins	Play-based learning activities	Focus on play, social development, less on subject-specific learning	Play-based learning activities	Two opposing fractions: the majority focus on play and social pedagogy and many of them look at early mathematics education with trepidation, but a growing group sees its importance and wants a stronger focus on learning activities	N.A.

enthusiasm for mathematics is related to enthusiasm for fostering mathematical skills in children (Vogt et al., 2022), which, in turn, influences the way in which a teacher designs learning environments. Although emotions are largely individual in nature, positive emotions about mathematics seem to be trainable, however, if the preparation program combines practical and theoretical training units (Thiel, 2022).

That early childhood teachers are often anxious about mathematics and tend to have little interest in the subject may lead to further consequences because this can also affect other areas of knowledge, especially MPCK. The influence of MCK on MPCK is controversially debated and some researchers judge it to be limited (Bruns et al., 2021), which could be put down to the fact, however, that in many studies the underlying conception of MCK was science-oriented rather than practice-oriented. This implies that the conceptualization of specialized knowledge as it was developed for school-related fields of work may not be easily transferable but needs other operationalizations with a practical orientation that also take the application of implicit knowledge into account and include situation-specific aspects (e.g., Gasteiger, Bruns, Benz, Brunner, & Sprenger, 2020; Torbeyns, Demedts, & Depaepe, 2022).

This distinction between science-related and practice-related facets of MCK (Bruns et al., 2021) could serve as a useful starting point for further research. For example, it would be worth examining whether anxiety about mathematics of early childhood teachers (e.g., Jenßen, 2022) or their emotions about mathematics (Thiel, 2022) could be explained by the assumption that their notion of mathematics commonly relates to the academic discipline while they do not see the immediate practical relevance of mathematics in everyday life. Furthermore, research into such questions should not neglect that the cultural and societal context shapes the image and thus individual notions of mathematics as well (Blömeke, Hsieh, Kaiser, & Schmidt, 2014; Dunekacke et al., 2016).

Job titles and general conditions

Are early childhood teachers primarily educators or are they rather like schoolteachers? A glance at the articles in this volume reveals a wide variety of job titles: they are called "early education teacher", "early childhood teacher", "early childhood education and care teacher", "early childhood educator", "preschool teacher", "kindergarten educator", or "kindergarten teacher". They work in "kindergarten", in "preschool", in "early childhood education and care", or in "early childhood education and care institutions". In view of this variety of terms for designation and workplace, it remains unclear whether all of them refer to comparable fields of work and professions. Without clarification of this fundamental terminological question, it is not possible to define a profile of professional skills and knowledge that is intended to cover the duties and activities of a supposedly homogenous professional group.

The results of a written survey of the authors who contributed to this volume can shed some light on this question, at least as far as their countries are concerned (Table 13.1). The selection of the framework conditions was based on the study by Gasteiger et al. (2021).

This limited selection of countries already suffices to show the wide diversity of central framework conditions (Table 13.1). Whether kindergarten attendance is compulsory for children (e.g., Switzerland) or optional (e.g., USA, Belgium) is likely to have a significant impact on the education of early childhood teachers and thus on their professional skills. The same applies to the age of the children and to whether there is a compulsory curriculum for early mathematics education. Furthermore, in some countries such as Australia or Germany, there are also marked regional differences in terms of general conditions, education, and kindergarten management that point to the heterogeneity within this only apparently homogenous professional group.

Education and professional development of early childhood teachers

As Table 13.1 indicates, there are also considerable differences with respect to the training of early childhood teachers. While some acquire a bachelor's degree, others complete their basic education at a vocational school. The structure and the academic level of the training, in turn, may lead to different expectations regarding performance (e.g., Rettenbacher, Eichen, Pfiffner, & Walter-Laager, 2022). The majority of early childhood teachers in all of the countries are trained as generalists. Whether and to what extent they acquire mathematics-specific content knowledge and pedagogical skills during their education varies greatly. As regards those training programs that explicitly include topics relating to mathematics, it would be worth investigating whether their conception of MCK is scientific or practical (Bruns et al., 2021).

According to the principle of life-long learning, professional skills are not only built up during the initial phase of training but constantly need to be developed throughout the professional career. For this reason, opportunities for further and continuing education that allow practicing early childhood teacher to expand and refine their professional skills are also important with respect to their qualification (Torbeyns et al., 2022, in this volume). This is reflected in the wide variety of concepts for professional development strategies, activities, and tools that are currently available (e.g., Gasteiger & Benz, 2018b; Sarama et al., 2022) and have proven to have positive effects on the early childhood teachers' skills (Bruns, Eichen, & Gasteiger, 2017; Gasteiger & Benz, 2018b; Sarama et al., 2022). The model of professional development along learning trajectories proposed by Sarama et al. (2022) as an example of such a program or course addresses different skills concerning understanding: understanding of goals, of children's thinking, and of effective teaching. In contrast, Gasteiger and Benz (2018b, pp. 13–14) devised a three-phase model for continuing education that addresses not only different skills but focuses on different "components of knowledge" depending on the phase. In Phase 1, explicit knowledge is at the center, and which can be taught. Phase 2 consists in "exploring mathematics in join-in studio" and focuses on "situational observing and perceiving" and pedagogical action, which is tested and practiced, and which focuses on gaining experience. This phase also pays attention to implicit knowledge. The final phase, Phase 3, is designed as a reflection meeting that reviews and evaluates the entire learning and training process.

That the focus on different areas of knowledge could be useful to foster the early childhood teachers' skills and may vary depending on the stage of the career can also be inferred from the results of the study by Dunekacke & Blömeke (2022). What contents and which knowledge components the training unit or professional-development activity should center on depends on the participants' prior knowledge and is thus likewise subject to cultural and societal framework conditions that structure and limit the education of early childhood teachers.

Conclusion

Do the foregoing sections provide a basis for answering the question concerning the professional skills that early childhood teachers need to be able to act competently? And can the answer take a general form when both the field of work and the training of early childhood teachers at least partly depend on cultural and societal norms and vary across countries and sometimes even within a country?

It is undisputed that early childhood teachers need explicit professional knowledge (Gasteiger & Benz, 2018a, 2018b). In the model proposed by Blömeke et al. (2015), professional knowledge is summarized, complemented by motivational-affective components and incorporated into the domain of dispositions. Other types of knowledge such as "interaction knowledge" or "counseling knowledge" (Baumert & Kunter, 2013) are considered to be relevant as well. Lindmeier (2011) subsumes these components under the basic skills and personal characteristics that are requisite for professional action. Depending on the research question, it makes sense, however, to subdivide professional knowledge into different areas of declarative knowledge (Anderson & Krathwohl, 2001; Dunekacke & Blömeke, 2022)—in the large three areas GPK, MCK, MPCK (Shulman, 1986) or in their further defined components (e.g., Ball, Thames, & Phelps, 2008) or in individual facets—and to operationalize them specifically for the field of work to which they apply. Explicit professional knowledge should be linked to implicit knowledge (Gasteiger & Benz, 2018a) and to appropriate experiences, as we see in models of teacher education and professional development (e.g., Gasteiger & Benz, 2018a; Sarama et al., 2022).

The model of transformation of professional competence (Blömeke et al., 2015), which structures this volume, does not primarily describe the individual areas of professional competence but rather the process of transforming professional knowledge areas, facets and dispositions into visible action in the professional field in the sense of performance. Lindmeier (2011) speaks in her three-component model of "action-related competencies", which develop alongside reflection skills on the basis of "basic knowledge". Planning and preparatory activities require reflection skills while action-related skills are needed in the enactment of the lesson and during the immediate pedagogical action in the situation. Dunekacke and Blömeke (2022, p. 121) understand the planning of mathematics-related actions ("ACT") as a "situation-specific skill", while Lindmeier (2011) considers preparatory planning to form part of a teacher's "reflective competence" and interprets immediate action in the situation of teaching as "action-related competence". Depending on the type of planning, in the situation or in advance, one or the other perspective might prove to be more adequate.

In the model by Blömeke et al. (2015), these action-related skills are interpreted as "situation-specific skills" and further subdivided into "perception", "interpretation", and "decision-making". In their entirety, these components lead to "performance" that manifests itself in observable behavior. In this model, "perception" ("PERC" in Dunekacke & Blömeke, 2022), which depends on professional knowledge, is of particular importance to the connection with situation-specific skills (e.g., Bruns et al., 2020; Dunekacke, 2016).

Do early childhood teachers in the field of early mathematics education therefore merely need—as suggested by the model by Blömeke et al. (2015)—appropriate dispositions and situation-specific skills? As mentioned, the field of work of early childhood teachers seems to differ significantly from that of schoolteachers in various respects and across national contexts. This leads to the question as to whether it is sufficient to model the requisite professional skills of early childhood teachers in accordance with common models. The model proposed by Gasteiger and Benz (2018a), which distinguishes between "explicit knowledge" and "implicit knowledge" and emphasizes individual diagnostic skills and the ability to support learning processes, at least suggests that models that only include the level of explicit knowledge might not suffice to capture the professional skills of early childhood teachers in an adequate way.

One open question that is in need of clarification concerns the process of the generation of visible performance in practice and the importance of different areas of knowledge in the modeling of professional action. In which of them is situation-related action grounded, both in prepared situations and immediately in the situation under time pressure? Furthermore, it needs to be clarified what role procedural knowledge plays in this transformation process and to what extent this component of explicit knowledge is accessible to early childhood teachers. Finally, it should be considered whether the model of transforming professional competence (Blömeke et al., 2015) could be extended by a fourth phase that consists in the monitoring or at least in the evaluation of one's professional actions as in the model of Gasteiger and Benz (2018a, 2018b).

This suggestion builds on the assumption that the ability to review, evaluate, and reflect on one's own performance in a specific pedagogical situation is indispensable with respect to effective professional development, on the one hand in the sense of the figure of the reflected practitioner (Schön, 1983) and on the other hand as an expression of professionalism. Such an extension would imply the necessity of enriching existing models by the aspect of professional self-regulation, which Baumert and Kunter (2013) regard as one of four aspects (convictions/values/goals, motivational orientation, self-regulation, and professional knowledge) of professional competence.

For the purpose of devising an all-including, holistic model that is explicitly tailored to early childhood teachers, the model of the transformation of professional competence by Blömeke et al. (2015) could therefore be combined with models of professional competence that have been specifically developed for this professional group (Gasteiger & Benz, 2018a, 2018b) and with models of basic professional knowledge components such as the one proposed by Lindmeier (2011) (Figure 13.2). Such a conceptual synthesis would take the different types of knowledge into account that are used in different ways in the individual phases of the transformation process of professional skills in more detail.

Both dispositions and situation-specific skills can be discussed against the background of explicit and implicit knowledge. Dispositions correspond to the concept of basic knowledge as proposed by Lindmeier (2011) (adapted for primary school teachers' knowledge by Knievel), which contains both explicit

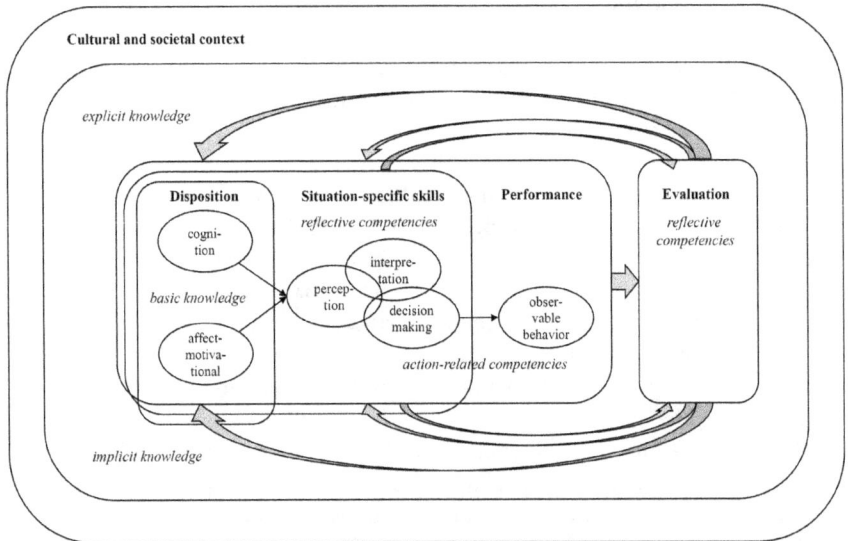

Figure 13.2 Conceptual synthesis for modeling the transformation of professional skills of early childhood teachers (based on Blömeke et al., 2015; Gasteiger & Benz, 2018a; Lindmeier, 2011).

and implicit components. Explicit as well as implicit knowledge is also included in situation-specific skills. Reflection skills are required when early childhood teachers plan a concrete teaching situation. Moreover, these skills are also relevant when, in connection with situation-specific skills, the action and/or the decision-making process are preceded by a brief evaluation. This would manifest itself in a minimal pause between perception and interpretation on the one hand and decision making on the other hand. The faster the decision making follows, the more likely it is that an early childhood teacher builds on implicit knowledge. Situation-specific skills are only transformed into performance and observable behavior, however, if action-related skills—also fed by implicit and explicit knowledge—are available and applied.

Although models that are specifically tailored to the professional profile of early childhood teachers could contribute to an improved understanding of teaching practices in this field, it should not be forgotten that both explicit and implicit knowledge are highly dependent on cultural and societal framework conditions and relate to a comparatively little standardized and regulated field of work. Therefore, it would be important to include central contextual factors systematically in future research on professional skills of early childhood teachers and to take them into account when interpreting the results. If the situation and the context of early childhood teachers' professional field of work differ from each other in terms of cultural patterns and influences, then context should be given increased attention—in research, in practice, and in the international discourse. Only an elucidation of the contextual factors allows an adequate interpretation of research findings. In order to examine the role and influences of the context, more cross-cultural studies are needed that compare the professional knowledge of early childhood teachers, the development of the children's performance, or the teaching practice in a detailed way. It is in this respect too that volumes that bring together comparative studies or studies from different countries and provide a solid basis for discussion are indispensable.

Notes

1 Occasionally, these models have also been referred to in the English-speaking discourse (e.g., Brühwiler & Blatchford, 2011).
2 The original model proposed by Lindmeier (2011) has been adapted to the professional competences of elementary mathematics teachers by Knievel, Lindmeier, and Heinze (2015) and to early childhood educators by Lindmeier et al. (2020).

References

Anderson, L. W., & Krathwohl, D. R. (2001). *A taxonomy for learning, teaching, and assessing: A revision of Bloom's taxonomy of educational objectives*. New York, NY: Longman.
Ball, D. L., Thames, M. H., & Phelps, G. (2008). Content knowledge for teaching: What makes it special? *Journal of Teacher Education, 59*(5), 389–407.

Baumert, J., & Kunter, M. (2013). The COACTIV model of teachers' professional competence. In M. Kunter, J. Baumert, W. Blum, U. Klusmann, S. Krauss, & M. Neubrand (Eds.), *Cognitive activation in the mathematics classroom and professional competence of teachers. Results from the COACTIV project* (pp. 25–48). New York, NY: Springer.

Björklund, C., & Palmer, H. (2022). Preschool teachers' ways of promoting mathematical learning in picture book reading. In S. Dunekacke, A. Jegodtka, T. Koinzer, K. Eilerts, & L. Jenßen (Eds.), *Early childhood teachers' professional competence in mathematics*. London: Routledge.

Blömeke, S., Gustafsson, J.-E., & Shavelson, R. J. (2015). Beyond dichotomies. *Zeitschrift für Psychologie, 223*(1), 3–13.

Blömeke, S., Hsieh, F.-J., Kaiser, G., & Schmidt, W. H. (2014). *International perspectives on teacher knowledge, beliefs and opportunities to learn. TEDS-M results*. Heidelberg: Springer.

Bruce, T. (1991). *Time to play in early childhood education*. Sevenoaks: Edward Arnold, Hodder & Stoughton.

Brühwiler, C., & Blatchford, P. (2011). Effects of class size and adaptive teaching competency on classroom processes and academic outcome. *Learning and Instruction, 21*, 95–108.

Bruns, J., Eichen, L., & Gasteiger, H. (2017). Mathematics-related competence of early childhood teachers visiting a continuous professional development course: An intervention study. *Mathematics Teacher Education and Development, 19*(3), 76–93.

Bruns, J., Gasteiger, H., & Strahl, C. (2021). Conceptualizing and measuring domain-specific content knowledge of early childhood educators: A systematic review. *Review of Education*. https://doi.org/10.1002/rev3.3255

Bruns, J., Strahl, C., & Gasteiger, H. (2020). Situative Beobachtung und Wahrnehmung frühpädagogischer Fachpersonenkräfte—Zum theoretischen Konstrukt und seiner empirischen Messung [Situational observation and perception of pre-service early childhood educators in the field of mathematics—Development and validation of a test instrument]. *Unterrichtswissenschaft*. https://doi.org/10.1007/s42010-020-00091-7

Burton, L. (2009). The culture of mathematics and the mathematical culture. In O. Skovsmose, P. Valero, & O. R. Christensen (Eds.), *University science and mathematics education in transition* (pp. 157–173). Boston, MA: Springer US.

Chen, J.-Q., McCray, J., Adams, M., & Leow, C. (2014). A survey study of early childhood teachers' beliefs and confidence about teaching early math. *Early Childhood Education Journal, 42*(6), 367–377.

Chiatovich, T., & Stipek, D. (2016). Instructional approaches in kindergarten. *The Elementary School Journal, 117*(1), 1–29.

Ciolan, L. E. (2013). Play to learn, learn to play: Creating better opportunities for learning in early childhood. *Procedia—Social and Behavioral Sciences, 76*, 186–189.

Clements, D., Sarama, J., & DiBiase, A. (2004). *Engaging young children in mathematics: Standards for early childhood mathematics*. Mahwah, NJ: Erlbaum.

Collins, A., Brown, J., & Newman, S. (1989). Cognitive apprenticeship: Teaching the crafts of reading, writing, and mathematics. In L. B. Resnick (Ed.), *Knowing, learning, and instruction: Essays in the honour of Robert Glaser* (pp. 453–495). Hillsdale, NJ: Erlbaum.

Common Core State Standards Initiative (2012). *Mathematics standards*. Retrieved 12 November 2020: Common Core State Standards Initiative website: http://www.core-standards.org/Math

Cooke, A., & Jay, J. (2022). Supporting preservice early childhood educators to identify mathematical activities in the actions of preverbal young children. In S. Dunekacke, A. Jegodtka, T. Koinzer, K. Eilerts, & L. Jenßen (Eds.), *Early childhood teachers' professional competence in mathematics*. London: Routledge.

De Haan, A. K. E., Elbers, E., & Leseman, P. P. M. (2014). Teacher- and child-managed academic activities in preschool and kindergarten and their influence on children's gains in emergent academic skills. *Journal of Research in Childhood Education, 28*(1), 43–58.

D-EDK (2014). *Lehrplan 21. Mathematik*. Bern: Projekt Lehrplan 21.

Dunekacke, S. (2016). *Mathematische Bildung in Alltags- und Spielsituationen begleiten— Handlungsnahe Erfassung mathematikdidaktischer Kompetenz angehender frühpädagogischer Fachkräfte durch die Bearbeitung von Videovignetten*. Berlin: Humboldt Universität zu Berlin.

Dunekacke, S., & Blömeke, S. (2022). Early mathematics education—What do student teachers learn? In S. Dunekacke, A. Jegodtka, T. Koinzer, K. Eilerts, & L. Jenßen (Eds.), *Early childhood teachers' professional competence in mathematics*. London: Routledge.

Dunekacke, S., Jenßen, L., Eilerts, K., & Blömeke, S. (2016). Epistemological beliefs of prospective preschool teachers and their relation to knowledge, perception, and planning abilities in the field of mathematics: A process model. *ZDM Mathematics Education, 48*(1–2), 125–137.

Fend, H. (1998). *Qualität im Bildungswesen: Schulforschung zu Systembedingungen, Schulprofilen und Lehrerleistung*. Weinheim: Juventa.

Gasteiger, H. (2012). Fostering early mathematical competencies in natural learning situations—Foundation and challenges of a competence-oriented concept of mathematics education in kindergarten. *Journal für Mathematik-Didaktik JMD, 33*(2), 181–201.

Gasteiger, H., & Benz, C. (2018a). Mathematics education competence of professionals in early childhood education—A theory-based competence model. In C. Benz, A. S. Steinweg, H. Gasteiger, P. Schöner, H. Vollmuth, & J. Zöllner (Eds.), *Mathematics education in the early years—Results from the POEM3 Conference, 2016* (pp. 69–91). Cham: Springer.

Gasteiger, H., & Benz, C. (2018b). Enhancing and analyzing kindergarten teachers' professional knowledge for early mathematics education. *Journal of Mathematical Behavior, 51*, 109–117.

Gasteiger, H., Brunner, E., & Chen, Ch.-S. (2021). Basic conditions of early mathematics education—A comparison between Germany, Taiwan and Switzerland. *International Journal of Science and Mathematics Education, 19*, 111–127.

Gasteiger, H., Bruns, J., Benz, C., Brunner, E., & Sprenger, P. (2020). Mathematical pedagogical content knowledge of early childhood teachers: A standardized situation-related measurement approach. *ZDM Mathematics Education, 52*(2), 193–205.

Giaquinto, M., Mancosu, P., Jorgensen, K. F., & Pedersen, S. A. (2005). Mathematical activity. In P. Mancosu, K. F. Jørgensen, & S. A. Pedersen (Eds.), *Visualization, Explanation and Reasoning Styles in Mathematics* (pp. 75–87). Dodrecht: Springer.

Ginsburg, H. P. (2022). Voices of competence: What I learned from my early education students. In S. Dunekacke, A. Jegodtka, T. Koinzer, K. Eilerts, & L. Jenßen (Eds.), *Early childhood teachers' professional competence in mathematics*. London: Routledge.

Hammer, M., & He, M. (2016). Preschool teachers' approaches to science: A comparison of a Chinese and a Norwegian kindergarten. *European Early Childhood Education Research Journal, 24*(3), 450–464.

Jegodtka, A., Hosoya, G., Szczesny, M., Jenßen, L., & Schmude, C. (2022). The development of children's mathematical skills and the quality of ECEC institutions. In S. Dunekacke, A. Jegodtka, T. Koinzer, K. Eilerts, & L. Jenßen (Eds.), *Early childhood teachers' professional competence in mathematics*. London: Routledge.

Jenßen, L. (2022). A math-avoidant profession: Review of the current research about early childhood teachers' mathematics anxiety and empirical evidence. In S. Dunekacke, A. Jegodtka, T. Koinzer, K. Eilerts, & L. Jenßen (Eds.), *Early childhood teachers' professional competence in mathematics*. London: Routledge.

KMK (2005). *Bildungsstandards der Kultusministerkonferenz. Erläuterungen zur Konzeption und Entwicklung*. München: Luchterhand.

Knievel, I., Lindmeier, A. M., & Heinze, A. (2015). Beyond knowledge: Measuring primary teachers' subject-specific competences in and for teaching mathematics with items based on video vignettes. *International Journal of Science and Mathematics Education, 13*(2), 309–329.

Krajewski, K., & Schneider, W. (2009). Early development of quantity to number-word linkage as a precursor of mathematical school achievement and mathematical difficulties: Findings from a four-year longitudinal study. *Learning and Instruction, 19*(6), 513–526.

Krajewski, K., Nieding, F., & Schneider, K. (2007). *Mengen, zählen, Zahlen*. Berlin: Cornelsen.

Lindmeier, A. (2011). *Modeling and measuring knowledge and competencies of teachers: A threefold domain-specific structure model for mathematics*. Münster: Waxmann.

Lindmeier, A., Seemann, S., Kuratli-Geeler, S., Wullschleger, A., Dunekacke, S., Leuchter, M., Vogt, F., Moser Opitz, E., & Heinze, A. (2020). Modelling early childhood teachers' mathematics-specific professional competence and its differential growth through professional development—An aspect of structural validity. *Research in Mathematics Education, 22*(2), 168–187.

NCTM. (2000). *Principles and standards for school mathematics*. Reston, VA: NCTM.

Oberhuemer, P. (2005). International perspectives on early childhood curricula. *International Journal of Early Childhood, 37*(1), 27–37.

OECD. (2011). *Starting strong III. A quality toolbox for early childhood education and care*. Paris: OECD Publishing.

Piaget, J. (1975). *Nachahmung, Spiel und Traum*. Stuttgart: Klett.

Pohle, L., Jenßen, L., & Eilerts, K. (2022). Early childhood teachers' selection of sub-skills-related activities and instructional approaches to foster children's early number skills. In S. Dunekacke, A. Jegodtka, T. Koinzer, K. Eilerts, & L. Jenßen (Eds.), *Early childhood teachers' professional competence in mathematics*. London: Routledge.

Presmeg, N. (2014). A dance of instruction with construction in mathematics. In U. Kortenkamp, B. Brandt, C. Benz, G. Krummheuer, S. Ladel, & R. Vogel (Eds.), *Early mathematics learning. Selected papers of the POEM 2012 Conference* (pp. 9–17). New York, NY: Springer.

Rettenbacher, K., Eichen, L., Pfiffner, M., & Walter-Laager, C. (2022). Age-appropriate performance expectations and learning objectives of early childhood teachers in the field of mathematics: a cross-country comparison of Austria and Switzerland. In S. Dunekacke, A. Jegodtka, T. Konizer, K. Eilerts, & L. Jenßen (Eds.), *Early childhood teachers' professional competence in mathematics*. London: Routledge.

Reusser, K., & Pauli, C. (2010). Unterrichtsgestaltung und Unterrichtsqualität—Ergebnisse einer internationalen und schweizerischen Videostudie zum Mathematikunterricht: Einleitung und Überblick. In K. Reusser, C. Pauli, & M. Waldis (Eds.), *Unterrichtsgestaltung und Unterrichtsqualität* (pp. 9–32). Münster: Waxmann.

Rezat, S. (2013). The textbook-in-use: Students' utilization schemes of mathematics text-books related to self-regulated practicing. *ZDM Mathematics Education, 45*(5), 659–670.

Rimm-Kaufman, S., & Sandilos, L. (2017). School transition and school readiness: An outcome of early childhood development. *Encyclopedia on early childhood development.* Retrieved on 13 February 2021 from https://www.researchgate.net/profile/Sara_Rimm-Kaufman/publication/251888181_School_Transition_and_School_Readiness_An_Outcome_of_Early_Childhood_Development/links/54f923b90cf2ccffe9e00740/School-Transition-and-School-Readiness-An-Outcome-of-Early-Childhood-Development.pdf

Rubin, K., & Pepler, D. J. (1982). Children's play: Piaget's views reconsidered. *Contemporary Educational Psychology, 7*(3), 289–299.

Sarama, J., & Clements, D. H. (2009). *Early childhood mathematics education research: Learning trajectories for young children.* New York: Routledge.

Sarama, J., Clements, D. H., & Stark Guss, S. (2022). Longitudinal evaluation of a scale-up modell for professional development in early mathematics. In S. Dunekacke, A. Jegodtka, T. Koinzer, K. Eilerts, & L. Jenßen (Eds.), *Early childhood teachers' professional competence in mathematics.* London: Routledge.

Schön, D. (1983). *The reflective practitioner: How professionals think in action.* New York, NY: Basic Books.

Seidel, T. (2014). Angebots-Nutzungs-Modelle in der Unterrichtspsychologie: Integration von Struktur- und Prozessparadigma. *Zeitschrift für Pädagogik, 60*(6), 850–866.

Seidel, T. (2020). Kommentar zum Themenblock "Angebots-Nutzungs-Modelle als Rahmung": Quo vadis deutsche Unterrichtsforschung? Modellierung von Angebot und Nutzung im Unterricht. *Zeitschrift für Pädagogik, 66*(Beiheft), 95–101.

Shulman, L. S. (1986). Those who understand: Knowledge growth in teaching. *Educational Researcher, 15*(2), 4–14.

Stebler, R., Vogt, F., Wolf, I., Hauser, B., & Rechsteiner, K. (2013). Play-based mathematics in kindergarten: A video analysis of children's mathematical behaviour while playing a board game in small groups. *Journal für Mathematik-Didaktik JMD, 34*(2), 149–175.

Thiel, O. (2022). How preservice teacher training chances prospective ECEC teachers' emotions about mathematics. In S. Dunekacke, A. Jegodtka, T. Koinzer, K. Eilerts, & L. Jenßen (Eds.), *Early childhood teachers' professional competence in mathematics.* London: Routledge.

Thiel, O., & Jenßen, L. (2018). Affective-motivational aspects of early childhood teacher students' knowledge about mathematics. *European Early Childhood Education Research Journal, 26*(4), 512–534.

Torbeyns, J., Demedts, F., & Depaepe, F. (2022). Preschool teachers' mathematical pedagogical knowledge and self-reported classroom activities. In S. Dunekacke, A. Jegodtka, T. Koinzer, K. Eilerts, & L. Jenßen (Eds.), *Early childhood teachers' professional competence in mathematics.* London: Routledge.

Valverde, G. A., Bianchi, L. J., Wolfe, R. G., Schmidt, W. H., & Houang, R. T. (2002). *According to the book—Using TIMSS to investigate the translation of policy into practice through the world of textbooks.* Dordrecht: Kluwer.

van den Boogaard, S., Van den Heuvel-Panhuizen, M., & Scherer, P. (2007). Picture book as a prompt for mathematical thinking by kindergartners: When Gaby was read "Being fifth". In Gesellschaft der Didaktik der Mathematik (Eds.), *Beiträge zum Mathematikunterricht 2007, 41. Jahrestagung der Gesellschaft für Didaktik der Mathematik vom 25.3. bis 30.3.2007 in Berlin* (pp. 831–834). Hildesheim: Franzbecker.

van den Heuvel-Panhuizen, M., & van den Boogaard, S. (2008). Picture books as an impetus for kindergartners' mathematical thinking. *Mathematical Thinking and Learning*, *10*(4), 341–373.

van Oers, B. (2010). Emergent mathematical thinking in the context of play. *Educational Studies of Mathematics*, *74*(1), 23–37.

Vogt, F., Leuchter, M., Dunekacke, S., Heinze, A., Lindmeier, A., Kuratli Geeler, S., Meier, A., Seemann, S., Wullschleger, A., & Moser Opitz, E. (2022). Kindergarten educators' affective-motivational dispositions: examining enthusiasm for fostering mathematics in kindergarten. In S. Dunekacke, A. Jegodtka, T. Koinzer, K. Eilerts, & L. Jenßen (Eds.), *Early childhood teachers' professional competence in mathematics*. London: Routledge.

Wahl, D. (1991). *Handeln unter Druck. Der weite Weg vom Wissen zum Handeln bei Lehrern, Hochschullehrern und Erwachsenenbildner* (2nd ed.). Weinheim: Deutscher Studien-Verlag.

Wittmann, E. C., & Müller, G. N. (2010). *Das Zahlenbuch. Begleitband zur Frühförderung*. Zug: Klett.

Index

.

For Product Safety Concerns and Information please contact our EU
representative GPSR@taylorandfrancis.com
Taylor & Francis Verlag GmbH, Kaufingerstraße 24, 80331 München, Germany

www.ingramcontent.com/pod-product-compliance
Lightning Source LLC
Chambersburg PA
CBHW060249220326
41598CB00027B/4036